Lecture Notes in Mathematics

Edited by A. Dold an

437

David Masser

Elliptic Functions and Transcendence

Springer-Verlag
Berlin · Heidelberg · New York 1975

Dr. D. W. Masser
Dept. of Mathematics
University of Nottingham
University Park
Nottingham NG7 2RD/England

Library of Congress Cataloging in Publication Data

Masser, David William, 1948-
Elliptic functions and transcendence.

(Lecture notes in mathematics ; 437)
Bibliography: p.
Includes index.
1. Functions, Elliptic. 2. Numbers, Transcendental.
I. Title. II. Series: Lecture notes in mathematics
(Berlin) ; 437.
QA3.L28 no. 437 [QA343] 510'.8s [515'.353]
74-32365

AMS Subject Classifications (1970): 10D25, 10F35, 33A25

ISBN 3-540-07136-9 Springer-Verlag Berlin · Heidelberg · New York
ISBN 0-387-07136-9 Springer-Verlag New York · Heidelberg · Berlin

Offsetdruck: Julius Beltz, Hemsbach/Bergstr.

CONTENTS

INTRODUCTION

These Notes are concerned with some new transcendence properties, or more exactly linear independence properties, of certain numbers associated with elliptic functions. The purpose of this general introduction is to describe the results and provide a historical context for them.

Essentially we treat special periodic functions $f(z)$ of a single complex variable z. The methods of transcendence theory are most likely to succeed when these functions have an algebraic addition theorem. Such functions are rather strictly delineated by an old theorem of Weierstrass, which says that up to algebraic dependence $f(z)$ must be one of the following.

First, $f(z)$ can be the exponential function $e^{\alpha z}$, and we take α as an algebraic number so that the differential equation of $f(z)$ is defined over the field $\mathbb{A}$ of algebraic numbers. Then it is a classical result of Lindemann that the fundamental period $2\pi i/\alpha$ of $f(z)$ is a transcendental number.

The second possibility is that $f(z)$ is a Weierstrass elliptic function $\wp(z)$ satisfying the differential equation

$$(\wp'(z))^2 = 4(\wp(z))^3 - g_2\wp(z) - g_3$$

for complex numbers g_2, g_3 with $g_2^3 \neq 27g_3^2$. Once again we take g_2, g_3 to be algebraic numbers (and this assumption will be maintained throughout these Notes). This is a doubly periodic function, and so we may choose a fundamental pair of periods ω_1, ω_2 with the imaginary part of ω_2/ω_1 positive. We also consider the associated Weierstrass zeta function $\zeta(z)$ defined by $\zeta'(z) = -\wp(z)$ and normalized additively to be an odd function of z. This possesses an adequate addition theorem, and there exist quasi-periods η_1 and η_2 corresponding to ω_1 and ω_2 such that

$$\zeta(z + \omega_i) = \zeta(z) + \eta_i \qquad (i = 1,2).$$

Siegel was the first to investigate the arithmetical properties of ω_1 and ω_2; he proved in [26] that they cannot both be algebraic numbers. A few years later Schneider in his fundamental researches improved on this result by obtaining the transcendence not only of ω_1 and ω_2 but also of η_1 and η_2. An account of this can be found in his book [25]. The exponential case above is naturally included by considering the five numbers ω_1, ω_2, η_1, η_2 and $2\pi i$ together; thus we see that each is transcendental. The most general theorem extending this statement would assert the algebraic independence of these five numbers, but there are two reasons why this cannot yet be proved. First, the general techniques are not available, and secondly, the Legendre relation

$$\omega_2\eta_1 - \omega_1\eta_2 = 2\pi i$$

shows that such a theorem would be false.

So we turn to the more fruitful question of linear independence: are the above five numbers together with 1 linearly independent over the field $\mathbb{A}$? The answer to this involves a splitting of cases. We consider the set of complex numbers $\lambda \neq 0$ such that $\wp(z)$ and $\wp(\lambda z)$ are algebraically dependent functions. By the addition theorem, this set together with $\lambda = 0$ forms a field $\mathbb{F}$. If $\mathbb{F} \neq \mathbb{Q}$ the field of rational numbers we say that $\wp(z)$ has complex multiplication; this is the exception rather than the rule and then $\mathbb{F}$ must be a complex quadratic extension $\mathbb{K}$ of $\mathbb{Q}$. Otherwise $\mathbb{F} = \mathbb{Q}$ means that $\wp(z)$ is without complex multiplication and in this case an affirmative answer to the above question had been conjectured.

Partial answers were first obtained by Schneider in [25], who proved that $2\pi i/\omega_1$, η_1/ω_1, and $\alpha\omega_1+\beta\eta_1$ are transcendental for non-zero algebraic numbers α and β. In addition he showed that ω_2/ω_1 is transcendental under the necessary condition that $\wp(z)$ has no complex multiplication; for otherwise ω_2/ω_1 lies in $\mathbb{K}$. These results were considerably generalized by Baker with the aid of his many-variable techniques. In two papers [2], [3] he proved the transcendence of any non-zero linear combination of ω_1, ω_2, η_1 and η_2 with algebraic coefficients, and shortly afterwards Coates [12] obtained the extended result for the five numbers ω_1, ω_2, η_1, η_2 and $2\pi i$. In Chapter II of these Notes we prove that these five numbers are themselves linearly independent when $\wp(z)$ has no complex multiplication; in conjunction with the result of Coates this establishes the linear independence of 1, ω_1, ω_2, η_1, η_2 and $2\pi i$ over $\mathbb{A}$. In other words, the vector space V spanned over $\mathbb{A}$ by these six numbers is of maximal dimension six.

The proof attempts to imitate the proof of the theorem of Coates mentioned above. The main difficulty arises from the periodicity of the auxiliary function Φ; this means that the zeros of Φ on the real diagonal do not carry as much information as they normally would. Therefore we are compelled to extrapolate further to deduce that Φ is small on a large part of the whole complex diagonal. After using a device to replace Φ by a simpler function Ψ, we may regard Ψ as a polynomial, and as such it inherits the smallness of Φ on a large subset $\mathcal{S}$ of $\mathbb{C}^3$. To show that $\mathcal{S}$ is well-distributed in a rather weak sense is a matter of diophantine approximation (this appears in a slightly disguised form in the proof); we then require arguments essentially in one complex variable to see that this impossibly restricts Ψ.

The results of Schneider mentioned above on the transcendence of $2\pi i/\omega_1$, η_1/ω_1 and ω_2/ω_1 can be regarded as first steps towards the theorem of Chapter II. The next (and most recent) step was taken by Coates who showed in [13] that ω_1, ω_2 and $2\pi i$ are linearly independent over $\mathbb{A}$ when $\wp(z)$ has no complex multiplication. The new idea in the proof of this result involves an appeal to the deep and extensive theory of Serre on division points of $\wp(z)$. However, for historical reasons the theory of transcendence has always laid emphasis on the concept of effectiveness; it requires, for example, that all constants occurring in the proofs should in principle be capable of explicit evaluation. Serre's theory was not constructed for transcendence purposes and sometimes fails to satisfy such criteria. Therefore it is of some interest to have a more elementary proof of Coates' result which does have an effective structure. We provide such a proof in Chapter IV.

So far we have not discussed what happens when $\wp(z)$ has complex multiplication. Since ω_2/ω_1 is an algebraic number it is clearly untrue that the six numbers 1, ω_1, ω_2, η_1, η_2 and $2\pi i$ are linearly independent over $\mathbb{A}$. In Chapter III we use a very simple argument to prove the slightly unexpected fact that there is another essentially distinct linear relation between the six numbers. This leaves four candidates for a basis of V, namely, 1, ω_1, η_1 and $2\pi i$, and in the same chapter we show that these four numbers are indeed linearly independent over $\mathbb{A}$. These results imply that the dimension of V is four in the case of complex multiplication.

After proving theorems of this type we consider a secondary objective of transcendence theory. This is to derive quantitative refinements in the form of measures; these tend to be more important for applications to other areas of number theory. In the situation outlined so far these refinements are of the following kind. If the coefficients α_0, α_1, α_2, β_1, β_2 and γ denote algebraic numbers not all zero, our results show that the expression

$$\Lambda = \alpha_0 + \alpha_1\omega_1 + \alpha_2\omega_2 + \beta_1\eta_1 + \beta_2\eta_2 + \gamma.2\pi i$$

does not vanish (under additional conditions depending on complex multiplication). The problem is then to find a positive lower bound for the absolute value of Λ in terms of the degrees and heights of the coefficients.

We shall give two examples of such estimates. In Chapter I we treat the case $\alpha_0 = \beta_1 = \beta_2 = \gamma = 0$ when there is no complex multiplication; this is equivalent to a transcendence measure for ω_2/ω_1. Feldman [16] has obtained a measure for a more general class of numbers but the lower bound turns out to be too weak for the application we have to make in Chapter II. Although our proof is just a modification of Feldman's argument*, it is important to

* Since I wrote these Notes, Feldman has published a paper in which he carries out this modification himself. This appears in Acta Arithmetica, 24, 477-489 and the result it contains is slightly sharper than mine.

give the details: first because of the crucial improvement in the lower bound, and secondly because all Feldman explicitly obtains in his paper is a measure of irrationality, which would be trivial for our particular number ω_2/ω_1 (since it is not real).

In our other example we consider the general form Λ when $\alpha_0 \neq 0$. Previously Baker [4] had given an estimate in the case $\beta_1 = \beta_2 = \gamma = 0$ of the form

$$|\Lambda| > C \exp(-(\log H)^{\kappa}),$$

where H is the maximum of the heights of α_0, α_1 and α_2, $C > 0$ depends only on their degrees and the numbers ω_1 and ω_2, and κ is a large absolute constant. In Chapter V we show that for the more general form κ can be taken as any number greater than 1 provided only $\alpha_0 \neq 0$; more precisely

$$|\Lambda| > C \exp(-\log H(\log\log H)^{7+\varepsilon}),$$

where H is now the maximum of the heights of α_0, α_1, α_2, β_1, β_2 and γ, and $C > 0$ depends only on their degrees, the numbers ω_1 and ω_2, and the arbitrarily chosen $\varepsilon > 0$. Here the dependence on H is quite near best possible, for standard arguments show that the absolute value of Λ can frequently be smaller than $H^{-\kappa}$ for some positive absolute constant κ.

This completes the outline of our study of the periods associated with $\wp(z)$. The remaining chapters investigate the arithmetical nature of a more general class of numbers, the algebraic points of $\wp(z)$. These are defined as complex numbers u such that either u is a pole of $\wp(z)$ or $\wp(u)$ is an algebraic number. The analogous definition for the exponential function e^z gives just the logarithms of algebraic numbers, and the theorem of Baker states that such numbers are linearly independent over $\mathbb{A}$ if they are linearly independent over $\mathbb{Q}$. We cannot enunciate the corresponding result for elliptic functions until we recall the definition above of the field $\mathbb{F}$ of complex multiplication; this is either $\mathbb{Q}$ or a complex quadratic extension $\mathbb{K}$ of $\mathbb{Q}$. Then it had been conjectured that algebraic points of $\wp(z)$ are linearly independent over $\mathbb{A}$ if they are linearly independent over $\mathbb{F}$.

Chapters VI and VII of these Notes are devoted to a proof of this conjecture when $\mathbb{F} \neq \mathbb{Q}$. In [25] Schneider obtained the first relevant result by proving the conjecture for two algebraic points and unrestricted $\mathbb{F}$. But for three or more algebraic points it seems necessary to take $\mathbb{F} \neq \mathbb{Q}$ in order that the techniques of Baker can be used. The advantage of complex multiplication is that the zeros of the auxiliary function can be taken on the diagonal corresponding to a part of the lattice of integers of $\mathbb{K}$. The extrapolation procedure operates on this wider range of zeros without the necessity of introducing division points, which would have involved algebraic number fields of intractably high degree. To obtain a final contradiction we use stronger versions of some of the ideas of Chapter II, but the proof is further complicated by the unruly behaviour of elliptic functions under differentiation. In fact we obtain a proof by induction on the number of algebraic points - not of the original conjecture but of a certain refinement. This gives a positive lower bound for the absolute value of the linear form

$$\Lambda = \alpha_1 u_1 + \ldots + \alpha_n u_n$$

where $u_1, \ldots, u_n$ are algebraic points linearly independent over $\mathbb{K}$ and $\alpha_1, \ldots, \alpha_n$ are algebraic numbers not all zero. More exactly, we show that for $\varepsilon > 0$ and a positive integer d there exists an effectively computable positive constant $C = C(\varepsilon, d, u_1, \ldots, u_n, g_2, g_3)$ such that

$$|\Lambda| > Ce^{-H^{\varepsilon}}$$

whenever the degrees and heights of $\alpha_1, \ldots, \alpha_n$ do not exceed d and H respectively. This compares favourably with the estimate of Feldman in [16] for $n = 2$ and unrestricted $\mathbb{F}$ which quantifies the result of Schneider mentioned above.

Such is the major part of these Notes; there remain the appendices. In the first of these we present a longer but more enlightening proof of a lemma appearing in Chapter III; it also gives a slightly better result. In Appendix II we consider two lemmas with a certain independent interest which were proved in Chapter II and Chapter VII. These concern the zeros of polynomials in several complex variables, and we use various methods to show that best possible estimates can sometimes be achieved.

Appendix III is an outline of how the proof in Chapter VII may be modified to yield a further transcendence result for algebraic points. The original proof gives a criterion for the possible vanishing of Λ; we now show that if Λ does not vanish it is a transcendental number. Together with the earlier result this implies that the numbers $1, u_1, \ldots, u_n$ are linearly independent over $\mathbb{A}$ if $u_1, \ldots, u_n$ are linearly independent over $\mathbb{K}$.

Finally Appendix IV deals with an application of the theorem of Chapter VII. If $\mathcal{C}$ is a curve of genus 1 defined over $\mathbb{Q}$, it is known that the magnitudes of rational points on $\mathcal{C}$ do not increase too rapidly in terms of their denominators. This does not appear to be explicitly stated in the literature, but it is implicit in the ideas of Siegel. It also follows from a result of Coates [11] giving a lower bound for the absolute value of Λ when $\alpha_1, \ldots, \alpha_n$ are rational integers; but the proof of this result uses the Thue-Siegel-Roth theorem and consequently shares the non-effective features of Siegel's work. When $\mathcal{C}$ has complex multiplication our theorem allows the estimate for rational points to be improved and made effective in terms of the Mordell-Weil group.

The research in these Notes was done at Trinity College, Cambridge with the support of a Science Research Council grant. I wish to thank Dr. J. Coates for generously making available to me some preliminary studies relating to the algebraic points problem of Chapter VII. In particular the credit for the main idea of Lemma 7.7 must be divided between him and my research supervisor Dr. A. Baker. But ultimately the work owes its existence to the advice and encouragement of Dr. Baker who first suggested the topic of the title as a possible field of research.

NOTATION

For a meromorphic function $f(z_1, \ldots, z_n)$ of the complex variables $z_1, \ldots, z_n$ we write
$f_{m_1,\ldots,m_n}(z_1, \ldots, z_n) = (\partial/\partial z_1)^{m_1} \ldots (\partial/\partial z_n)^{m_n} f(z_1, \ldots, z_n)$,
and for non-negative integers ℓ, m we denote by $\wp(z,\ell,m)$ the m-th derivative of $(\wp(z))^{\ell}$. If α is an algebraic number we define its size as the maximum of the absolute values of its conjugates and we say that an integer $a > 0$ is a denominator for α if $a\alpha$ is an algebraic integer. The height H of α is the maximum of the absolute values of the relatively prime integer coefficients in the minimal polynomial of α. Thus α has a denominator not exceeding H given by the leading coefficient of this polynomial. Also if d is the degree of α, the inequality $|\alpha| \leqslant dH$ is easily verified (see [1], p.206) and clearly the same estimate is valid for the size of α.

For a real number x we denote by $[x]$ the greatest integer not exceeding x.

Finally we use $c, c_1, \ldots$ to signify positive constants depending on various parameters, and we adopt the convention that the constants appearing in the proof of a lemma are allowed to depend only on those parameters appearing in its enunciation.

CHAPTER ONE

1.1 Introduction

In this chapter we shall establish a transcendence measure for the ratio $\tau = \omega_2/\omega_1$ when $\wp$ has no complex multiplication. Since the measure turns out to be stronger than that implied by the more general result [16] of Feldman, it has some independent interest; however, it will also be an essential component of the proof of Theorem II.

Theorem I

For any $\varepsilon > 0$ we have

$$|\tau - \alpha| > C \exp(-(\log H)^{3+\varepsilon}) \tag{1}$$

for all algebraic numbers α of height H, where $C > 0$ depends only on τ, ε and the degree of α.

1.2 Preliminary lemmas

In this section we collect together some elementary lemmas needed for the proof of Theorem I. The first two deal with simple properties of $\wp(z)$.

Lemma 1.1

There is a closed disc $\mathcal{D}$ in the complex plane with centre at $z = \frac{1}{4}$ such that

(i) $\wp(\omega_1 z)$, $\wp(\omega_2 z)$ are regular in $\mathcal{D}$

(ii) If z, z' lie in $\mathcal{D}$

$$c_1|z - z'| \leqslant |\wp(\omega_i z) - \wp(\omega_i z')| \leqslant c_2|z - z'|$$

for $i = 1,2$, where c_1, c_2 depend only on ω_1 and ω_2.

Proof

Since $\wp'(\frac{1}{4}\omega_i) \neq 0$, we can find a closed disc $\mathcal{D}$ centred at $z = \frac{1}{4}$ in which $\wp(\omega_1 z)$ and $\wp(\omega_2 z)$ are regular and simple (see for example [14] p.260). Then the continuous function

$$\phi(z,z') = (\wp(\omega_i z) - \wp(\omega_i z'))/(z - z') \qquad (z \neq z')$$

$$= \omega_i \wp'(\omega_i z) \qquad (z = z')$$

does not vanish on the compact set $\mathcal{D} \times \mathcal{D}$ and thus its absolute value possesses a positive lower bound and a finite upper bound in this region.

Lemma 1.2

For any positive integer ℓ, the m-th derivative of $(\wp(z))^{\ell}$ can be expressed in the form

$$\wp(z,\ell,m) = \sum U(t,t',t'',m,\ell)(\wp(z))^{t}(\wp'(z))^{t'}(\wp''(z))^{t''}$$

where the summation is over all non-negative integers t, t', t'' with

$$2t + 3t' + 4t'' = m + 2\ell$$

and $U(t,t',t'',m,\ell)$ denotes a rational integer with absolute value at most $m!c_3^{\ell+m}$ for some absolute constant c_3.

Proof

See Lemma 2 of [3].

Lemma 1.3

For an integer $L \geqslant 1$ and complex numbers $p(\lambda)$ $(0 \leqslant \lambda \leqslant L)$

let
$$\phi(z) = \sum_{\lambda=0}^{L} p(\lambda) z^{\lambda}.$$

Suppose $\sigma_0, \ldots, \sigma_L$ are distinct complex numbers with

$$\min_{\lambda \neq \mu} |\sigma_\lambda - \sigma_\mu| \geq \delta \ , \ \max_\lambda |\sigma_\lambda| \leq S$$

where $S \geq 1 \geq \delta$. Then for all λ we have

$$|p(\lambda)| \leq (c_4 S/\delta)^L \max_\mu |\phi(\sigma_\mu)|$$

for some absolute constant c_4.

Proof

We write

$$P(z) = \prod_{\lambda=0}^{L} (z - \sigma_\lambda)$$

and use the interpolation formula of Lagrange

$$\phi(z) = \sum_{\lambda=0}^{L} \phi(\sigma_\lambda) P(z)/(P'(\sigma_\lambda)(z - \sigma_\lambda))$$

to estimate the coefficients of $\phi(z)$, noting that $|P'(\sigma_\lambda)| \geq \delta^{-L}$ and also that the coefficients of $P(z)/(z - \sigma_\lambda)$ do not exceed $(c_5 S)^L$ in absolute value.

Now we let $\mathcal{E}$ be the set of points of the complex plane that are congruent to a point of $\mathfrak{D}$ modulo the periods of $\wp(\omega_1 z)$ and the periods of $\wp(\omega_2 z)$. The next lemma exhibits 'many' points of $\mathcal{E}$.

Lemma 1.4

There is a constant c_6 depending only on ω_1 and ω_2 such that for any integer $L > c_6$ the following is true. There are at least $L+1$ distinct integers $r_0, \ldots, r_L$ with absolute values at most $c_7 L$ such that

$$z(\ell, m) = \tfrac{1}{4} + \ell L^{-2} + r_m \tau \qquad (2)$$

lies in $\mathcal{E}$ for all integers ℓ, m with $0 \leq \ell, m \leq L$.

Proof

Let c be a constant depending only on ω_1 and ω_2 so large that the estimates below are valid. When A, B, C are non-negative integers not exceeding $T > c$, the points $A\omega_1^2 + B\omega_1\omega_2 + C\omega_2^2$ lie in a square centred at the origin with side at most c_8T. This square may be divided into at most $c_9(cT)^2$ smaller disjoint squares of side c^{-1}; hence if

$$\tfrac{1}{2}T^3 < c_9(L+1)c^2T^2 < T^3 \qquad (3)$$

at least one smaller square contains at least L+2 of these points. Then the L+1 differences between a fixed one and the others have absolute values at most $2c^{-1}$ and are of the form

$$\delta_m = p_m\omega_1^2 + q_m\omega_1\omega_2 + r_m\omega_2^2 \qquad (0 \leqslant m \leqslant L).$$

The resulting value of $z(\ell,m)$ given by (2) is clearly congruent to $\tfrac{1}{4} + \ell L^{-2}$ modulo the periods of $\wp(\omega_1 z)$. Also since

$$r_m\tau = -q_m - p_m\omega_1/\omega_2 + \delta_m/\omega_1\omega_2$$

we see that $z(\ell,m)$ is congruent to $\tfrac{1}{4} + \ell L^{-2} + \delta_m/\omega_1\omega_2$ modulo the periods of $\wp(\omega_2 z)$. Thus if $L > c$ the point $z(\ell,m)$ lies in $\mathcal{E}$ for $0 \leqslant \ell,m \leqslant L$.

Finally if $r_m = r_n$ for $m \neq n$ it would follow that

$$|p\omega_1^2 + q\omega_1\omega_2| \leqslant |\delta_m| + |\delta_n| \leqslant 4c^{-1}$$

where $p = p_m - p_n$, $q = q_m - q_n$. Since p,q are not both zero and ω_1/ω_2 is not real this is impossible for large enough c. Therefore $r_0, \ldots, r_L$ are distinct and from (3)

$$|r_m| \leqslant 2T < c_7L.$$

This completes the proof.

Lemma 1.5

There is a constant c_{10} depending only on ω_1 and ω_2 with the following property. For an integer $L \geqslant 1$ and complex numbers $p(\lambda_1,\lambda_2)$ $(0 \leqslant \lambda_1,\lambda_2 \leqslant L)$ let

$$\phi(z) = \sum_{\lambda_1=0}^{L} \sum_{\lambda_2=0}^{L} p(\lambda_1,\lambda_2)(\wp(\omega_1 z))^{\lambda_1}(\wp(\omega_2 z))^{\lambda_2}$$

and for $r \geqslant 0$ let $M(r)$ be the maximum modulus of $\phi(z)$ at points z of $\mathcal{E}$ with $|z| \leqslant r$. Then if

$$\mu = \min |A\omega_1^2 + B\omega_1\omega_2 + C\omega_2^2|$$

taken over all integers A, B, C, not all zero, with absolute values at most $c_{10}L$, we have for all λ_1,λ_2

$$|p(\lambda_1,\lambda_2)| \leqslant (L/\mu)^{c_{11}L} M(c_{12}L) .$$

Proof

As in the previous lemma we take $L > c$. From that lemma there are L+1 distinct integers $r_0, \ldots ,r_L$ of absolute value at most $c_{13}L$ such that the points

$$z(\ell,m) = \tfrac{1}{4} + \ell L^{-2} + r_m\tau \qquad (0 \leqslant \ell,m \leqslant L)$$

lie in $\mathcal{E}$. Hence $|z(\ell,m)| < c_{12}L$ and if $M = M(c_{12}L)$ we have

$$|\phi(z(\ell,m))| \leqslant M. \tag{4}$$

If we fix ℓ and write

$$q(\lambda_2) = \sum_{\lambda_1=0}^{L} p(\lambda_1,\lambda_2)(y(\ell))^{\lambda_1} \tag{5}$$

where

$$y(\ell) = \wp(\tfrac{1}{4}\omega_1 + \omega_1 \ell L^{-2})$$

it follows that

$$\phi(z(\ell,m)) = \sum_{\lambda_2=0}^{L} q(\lambda_2)(x(m))^{\lambda_2} \tag{6}$$

where

$$x(m) = \wp(\omega_2 z(\ell,m)).$$

We proceed to use Lemma 1.3 to derive upper bounds first for $|q(\lambda_2)|$ and then for $|p(\lambda_1,\lambda_2)|$. From Lemma 1.1 it is clear that if $m \neq n$

$$|x(m)| < c_{14} \quad , \quad |x(m) - x(n)| > c_{15}\xi ,$$

where

$$\xi = |(r_m - r_n)\tau - \Omega_2|$$

and Ω_2 is the pole of $\wp(\omega_2 z)$ nearest to $(r_m - r_n)\tau$. If $\Omega_2 = q + p\omega_1/\omega_2$ for integers p,q we must have $|p|,|q| < c_{16}L$ and so

$$\xi = |\omega_1\omega_2|^{-1}\,|p\omega_1^2 + q\omega_1\omega_2 - (r_m - r_n)\omega_2^2| > c_{17}\mu$$

provided c_{10} is large enough. Therefore from (6), (4) and Lemma 1.3 we have for all λ_2

$$|q(\lambda_2)| \leqslant (L/\mu)^{c_{18}L} M.$$

Finally it is easy to see that if $\ell \neq m$

$$|y(\ell)| < c_{19} , \quad |y(\ell) - y(m)| > L^{-c_{20}} ,$$

whence from (5) we obtain for all λ_1,λ_2

$$|p(\lambda_1,\lambda_2)| \leqslant (L/\mu)^{c_{21}L} M.$$

We conclude this section with two simple but useful algebraic lemmas.

Lemma 1.6

Let $\alpha_1, \ldots ,\alpha_n$ be algebraic numbers of degrees at most d and heights at most $H \geqslant 2$, and write $\mathbb{L} = \mathbb{Q}(\alpha_1, \ldots ,\alpha_n)$. Then there is an integral basis of $\mathbb{L}$ over $\mathbb{Q}$ consisting of numbers $w_1, \ldots ,w_N$ of size at most $H^{c_{22}}$. Furthermore if λ is an algebraic integer of $\mathbb{L}$ with size s then

$$\lambda = m_1 w_1 + \ldots + m_N w_N$$

for rational integers $m_1, \ldots, m_N$ with absolute values at most $H^{c_{22}}s$. Here c_{21} and c_{22} depend only on n and d.

Proof

Let $N = [\mathbb{L}:\mathbb{Q}]$ and denote by $\sigma_1, \ldots, \sigma_N$ the embeddings of $\mathbb{L}$ into $\mathbb{C}$. For integers $\ell_1, \ldots, \ell_n$ we write

$$\alpha = \ell_1 \alpha_1 + \ldots + \ell_n \alpha_n$$

and

$$\delta_{jk} = \sum_{i=1}^{n} \ell_i (\alpha_i^{\sigma_j} - \alpha_i^{\sigma_k}) \qquad (1 \leq j < k \leq N).$$

Now the number of integral points $(\ell_1, \ldots, \ell_n)$ in a real n-dimensional ball of radius R centred at the origin is at least $c_{23}R^n$, while the number of these on the hyperplane $\delta_{jk} = 0$ for some j,k is at most $\frac{1}{2}N(N-1)c_{24}R^{n-1}$. Hence by taking R large enough and noting that $N \leq d^n$ we can find integers $\ell_1, \ldots, \ell_n$ with absolute values at most c_{25} such that $\delta_{jk} \neq 0$ for all j,k. It is a well-known consequence of these inequalities that the corresponding algebraic number α generates $\mathbb{L}$. Further there is a positive integer a not exceeding H^n such that $\theta = a\alpha$ is an algebraic integer of size at most $H^{c_{26}}$. Hence if Δ denotes the rational integer

$$\Delta = \prod_{j \neq k} |\theta^{\sigma_j} - \theta^{\sigma_k}|$$

and Θ is the $\mathbb{Z}$-module with basis

$$\Delta^{-1}, \Delta^{-1}\theta, \ldots, \Delta^{-1}\theta^{N-1}$$

we have

$$\Delta\Theta \subseteq I \subseteq \Theta,$$

where I is the ring of integers of $\mathbb{L}$. Then the diagonal procedure (see [28], p.144) gives an integral basis for I of the form

$$w_i = c_{i1}\Delta^{-1} + c_{i2}\Delta^{-1}\theta + \dots + c_{ii}\Delta^{-1}\theta^{i-1} \qquad (1 \leqslant i \leqslant N)$$

where c_{ij} is a rational integer with $0 \leqslant c_{ij} \leqslant \Delta$. Hence the size of w_i is at most $NH^{c_{26}(N-1)} < H^{c_{21}}$. Also any algebraic integer λ of $\mathbb{L}$ may be written as

$$\lambda = m_1 w_1 + \dots + m_N w_N$$

for rational integers $m_1, \dots, m_N$, and by applying σ_i to both sides we obtain a set of linear equations for these coefficients. The determinant of these equations has absolute value at least unity, since it is the square root of the discriminant of $\mathbb{L}$. The asserted estimates for $m_1, \dots, m_N$ now follow at once, and this completes the proof of the lemma.

Lemma 1.7

Let M,N be integers with $N > M > 0$, and let u_{ij} $(1 \leqslant i \leqslant M,\ 1 \leqslant j \leqslant N)$ be rational integers with absolute values at most $U \geqslant 1$. Then there exist rational integers $x_1, \dots, x_N$, not all zero, with absolute values at most $(NU)^{M/(N-M)}$, such that

$$\sum_{j=1}^{N} u_{ij} x_j = 0 \qquad (1 \leqslant i \leqslant M).$$

Proof

This is a well-known result and a proof can be found in [1], p.208.

1.3 Proof of Theorem I

We start by assuming that for some ε with $0 < \varepsilon < \frac{1}{2}$ there exists an algebraic number α of degree d and height at most H such that

$$|\tau - \alpha| < \exp(-(\log H)^{3+\varepsilon}). \tag{7}$$

Further we suppose that if $d \leqslant 2$ the height of α is exactly H. We renumber constants afresh, and denote by $c, c_1, c_2, \ldots$ positive constants depending only on ω_1, ω_2, d and ε. Since $\wp(z)$ has no complex multiplication, (1) is trivial for $d \leqslant 2$ and $H \leqslant c$, and so Theorem I will be proved if we obtain a contradiction for $H > c$. Here c is sufficiently large for the validity of the subsequent estimates.

On setting

$$\delta = \varepsilon/100, \quad k = [(\log H)^{2+40\delta}], \quad L = [k^{\frac{1}{2}+\delta}]$$

the proof proceeds by a sequence of lemmas.

Lemma 1.8

There are rational integers $p(\lambda_1,\lambda_2)$, not all zero, with absolute values at most $H^{c_1 k}$, such that the function

$$\Phi(z) = \sum_{\lambda_1=0}^{L} \sum_{\lambda_2=0}^{L} p(\lambda_1,\lambda_2)(\wp(\omega_1 z))^{\lambda_1}(\wp(\omega_2 z))^{\lambda_2}$$

satisfies

$$|\Phi_m(\tfrac{1}{4})| < \exp(-k^{\frac{3}{2}+15\delta}) \tag{8}$$

for all integers m with $0 \leqslant m \leqslant k$.

Proof

We shall choose the integers $p(\lambda_1,\lambda_2)$ such that for all integers m with $0 \leqslant m \leqslant k$ we have $A_m = 0$, where

$$A_m = \sum_{\lambda_1=0}^{L} \sum_{\lambda_2=0}^{L} p(\lambda_1,\lambda_2)Q(\lambda_1,\lambda_2,m)$$

and

$$Q(\lambda_1,\lambda_2,m) = \sum_{\mu=0}^{m} \binom{m}{\mu} \alpha^{\mu} \wp(\tfrac{1}{4}\omega_1,\lambda_1,m-\mu)\,\wp(\tfrac{1}{4}\omega_2,\lambda_2,\mu).$$

From Lemma 1.2 we see that if $\lambda_i \leqslant L$ and $\mu_i \leqslant k$ the algebraic numbers $\wp(\tfrac{1}{4}\omega_i,\lambda_i,\mu_i)$ $(i = 1,2)$ have sizes at most

$$(\mu_i + 2\lambda_i + 1)^3 \mu_i!\, c_2^{\lambda_i+\mu_i} < k^{c_3 k} \tag{9}$$

and a common denominator p at most $c_4{}^k$. Hence if $a \leqslant H$ is a denominator for α the numbers $p^2 a^k Q(\lambda_1,\lambda_2,m)$ are algebraic integers of the field $\mathbb{F}$ generated over $\mathbb{Q}$ by the numbers

$$g_2,\ g_3,\ \alpha,\ \wp(\tfrac{1}{4}\omega_i),\ \wp'(\tfrac{1}{4}\omega_i),\ \wp''(\tfrac{1}{4}\omega_i) \qquad (i = 1,2)$$

and their sizes do not exceed

$$p^2 a^k (k + 1) 2^k (dH)^k k^{2c_3 k} < H^{c_5 k}$$

since $H > k$. Thus if $f = [\mathbb{F}:\mathbb{Q}]$ and $w_1, \ldots ,w_f$ is an integral basis chosen according to Lemma 1.6 we may write

$$p^2 a^k Q(\lambda_1,\lambda_2,m) = n_1 w_1 + \ldots + n_f w_f$$

where n_i denotes a rational integer with absolute value at most $H^{c_6 k}$. It follows that if the f equations

$$\sum_{\lambda_1=0}^{L} \sum_{\lambda_2=0}^{L} p(\lambda_1,\lambda_2) n_i = 0 \qquad (1 \leqslant i \leqslant f)$$

hold for all integers m with $0 \leqslant m \leqslant k$ we shall have $A_m = 0$. These are $M = (k + 1)f$ equations in $N = (L + 1)^2$ unknowns $p(\lambda_1,\lambda_2)$, and since

$$N > k^{1+\delta} > 2M$$

we may use Lemma 1.7 to find a non-trivial solution in rational integers with absolute values at most

$$N \max_{1\leqslant i\leqslant f} |n_i| < H^{c_7 k}.$$

It remains to verify the inequalities (8). But it is clear that $\omega_1^{-m}\Phi_m(\tfrac{1}{4})$ is the expression A_m with α replaced by

τ in $Q(\lambda_1,\lambda_2,m)$; since $\delta < 1/200$ we have

$$3 + 100\delta > (2 + 40\delta)(\tfrac{3}{2} + 18\delta)$$

and hence from (7)

$$|\tau - \alpha| < \exp(-k^{3/2+18\delta}).$$

Thus for $0 \leqslant \mu \leqslant k^{3/2}$ we see that

$$|\tau^{\mu} - \alpha^{\mu}| \leqslant c_7^{\mu}|\tau - \alpha| < \exp(-k^{3/2+17\delta})$$

and if $m \leqslant k^{3/2}$ the coefficients of α^{μ} in $Q(\lambda_1,\lambda_2,m)$ do not exceed $k^{c_8 k^{3/2}}$ in absolute value (cf. (9)); whence for all integers m with $0 \leqslant m \leqslant k^{3/2}$ we have

$$|\omega_1^{-m}\Phi_m(\tfrac{1}{4}) - A_m| < \exp(-k^{3/2+16\delta}). \tag{10}$$

The assertions (8) now follow.

Lemma 1.9

Suppose $Z \geqslant 6$ and let $\phi(z) = P(z)\Phi(z)$, where

$$P(z) = \prod_{i=1}^{2}\prod_{\Omega_i}(\omega_i z - \Omega_i)^{2L}$$

and Ω_i runs through all periods of $\wp(z)$ with $|\Omega_i| \leqslant |\omega_i|Z$ $(i = 1,2)$. Then $\phi(z)$ is regular for $|z| \leqslant Z$, and for any z with $|z| \leqslant \frac{1}{2}Z$ and any non-negative integer m we have

$$|\phi(z)| < H^{c_9 k} Z^{c_{10} L Z^2}, \qquad |P_m(z)| < m^m Z^{c_{11} L Z^2}.$$

Proof

The first inequality is clear from Lemma 1.8 and lemma 1 of [3], while the second is a consequence of the Cauchy integral

$$P_m(\zeta) = \frac{m!}{2\pi i}\int \frac{P(z)\,dz}{(z - \zeta)^{m+1}}$$

taken over the positively oriented unit circle with centre at $z = \zeta$.

Lemma 1.10

For all integers m with $0 \leqslant m \leqslant k^{1+6\delta}$ we have

$$|\Phi_m(\tfrac{1}{4})| < \exp(-k^{3/2+15\delta}).$$

Proof

Since $\Phi(z)$ has period 1 we see from Lemma 1.8 that for all integers s,m with

$$1 \leqslant s \leqslant h = [k^{1/2-2\delta}] \quad , \quad 0 \leqslant m \leqslant k \tag{11}$$

the inequality

$$|\Phi_m(s + \tfrac{1}{4})| < \exp(-k^{3/2+15\delta})$$

is valid. We set $Z = 10h$ and denote by $\phi(z)$ the corresponding function defined in Lemma 1.9. Further we write

$$F(z) = \prod_{s=1}^{h} (z - s - \tfrac{1}{4})^k.$$

Therefore for integers s,m in the ranges (11) we have

$$\phi_m(s + \tfrac{1}{4}) = \sum_{\mu=0}^{m} \binom{m}{\mu} P_\mu(s + \tfrac{1}{4})\Phi_{m-\mu}(s + \tfrac{1}{4}).$$

The right side of this may be estimated by means of Lemma 1.9; since

$$LZ^2 < c_{12}hk^{1-\delta} < k^{3/2} \tag{12}$$

we find that

$$|\phi_m(s + \tfrac{1}{4})| \leqslant k^{c_{13}k^{3/2}} \max_{0 \leqslant \mu \leqslant m} |\Phi_\mu(\tfrac{1}{4})| < \exp(-k^{3/2+14\delta}). \tag{13}$$

Let ζ be any number in the region $\mathfrak{D}$ of Lemma 1.1. We proceed to derive an upper bound for $|\phi(\zeta)|$ using the formula

$$\frac{\phi(\zeta)}{F(\zeta)} = \frac{1}{2\pi i} \int_C \frac{\phi(z)\, dz}{(z - \zeta)F(z)} - \frac{1}{2\pi i} \sum_{m=0}^{k} \sum_{s=1}^{h} \frac{\phi_m(s+\frac{1}{4})}{m!} I(m,s) \tag{14}$$

where C is the positively described circle $|z| = 5h$;

$$I(m,s) = \int_{C_s} \frac{(z - s - \frac{1}{4})^m}{(z - \zeta)F(z)}\, dz ,$$

and C_s is the positively described circle with centre at

$s+\frac{1}{4}$ and radius that of $\mathfrak{D}$. Note that since this radius cannot exceed $\frac{1}{4}$, the circles C_s do not intersect each other.

For z on C we have from Lemma 1.9

$$|\phi(z)| < H^{c_9 k} Z^{c_{10} L Z^2}$$

while for each linear factor of F(z)

$$|z - s - \tfrac{1}{4}| > 2|\zeta - s - \tfrac{1}{4}|$$

and from this last inequality

$$|F(\zeta)/F(z)| \leqslant 2^{-hk} .$$

Also, since $|F(z)| > 1$ for z on C_s, it is clear that

$$|I(m,s)| < c_{14}^{k}$$

whence from the estimates (12), (13) and

$$\log H < k^{\frac{1}{2}-9\delta} < hk^{-5\delta} \quad , \quad |F(\zeta)| < (2h)^{hk} < e^{k^{3/2}}$$

we deduce that

$$|\phi(\zeta)| < \exp(-c_{15}hk) .$$

Now the absolute value of each linear factor of $P(\zeta)$ exceeds 2 with at most $c_{16}L$ exceptions, and since ζ is in $\mathfrak{D}$ we have the lower bound c_{17} for these exceptions. Thus $|P(\zeta)| > 1$ and we conclude that

$$|\Phi(\zeta)| < \exp(-c_{15}hk) .$$

We now use Cauchy's integral

$$\Phi_m(\tfrac{1}{4}) = \frac{m!}{2\pi i} \int \frac{\Phi(z)\, dz}{(z - \frac{1}{4})^{m+1}}$$

taken over the positively described boundary of $\mathfrak{D}$ to deduce that if $0 \leqslant m \leqslant k_1 = [k^{1+b\delta}]$

$$|\Phi_m(\tfrac{1}{4})| < \exp(-c_{18}hk) .$$

Therefore from (10)

$$|A_m| < \exp(-c_{19}hk) .$$

On the other hand A_m is an algebraic number of degree at most c_{20}, and the estimates of Lemma 1.8 show its size to be at most $H^{c_{21}k_1}$; thus if $A_m \neq 0$ we obtain the lower bound $|A_m| > H^{-c_{22}k_1}$. Since

$$k_1 \log H < hk^{1-\delta}$$

this implies $A_m = 0$, and the inequality of the lemma now follows from (10).

Lemma 1.11

For all z in $\mathcal{E}$ with $|z| \leq k^{\frac{1}{2}+2\delta}$ we have

$$|\Phi(z)| < e^{-k^{3/2}}.$$

Proof

From the preceding lemma we see that

$$|\Phi_m(s + \tfrac{1}{4})| < \exp(-k^{3/2+15\delta})$$

for all integers s,m with

$$1 \leq s \leq h_1 = [k^{\frac{1}{2}+2\delta}] \quad , \quad 0 \leq m \leq k_1 . \tag{15}$$

We define the function $P(z)$ as before with h replaced by h_1, and since

$$Lh_1^2 < c_{23}h_1k_1k^{-2\delta} < k^{3/2+7\delta} \tag{16}$$

we deduce that (13) holds for the range (15). From the periodicity of $\Phi(z)$ it clearly suffices to prove the lemma for complex numbers ζ in $\mathcal{E}$ with

$$|\zeta| \leq k^{\frac{1}{2}+2\delta} \quad , \quad |\mathrm{Re}(\zeta)| \leq \tfrac{1}{2}.$$

We write

$$F(z) = \prod_{s=1}^{h_1} (z - s - \tfrac{1}{4})^{k_1}$$

and use (14) when C is the positively described circle $|z| = 5h_1$ and the sum runs over integers s,m in the ranges (15).

The additional restriction on ζ ensures that $|I(m,s)| < c_{24}^{k}$, and from (16) we eventually find that

$$|\phi(\zeta)| < \exp(-c_{25}h_1k_1).$$

As before $|P(\zeta)| > 1$, and therefore

$$|\Phi(\zeta)| < e^{-k^{3/2}}$$

which completes the proof of the lemma.

We now prove Theorem I as follows. From Lemma 1.5 we see that for all λ_1 , λ_2

$$|p(\lambda_1,\lambda_2)| \leqslant (L/\mu)^{c_{26}L} \; e^{-k^{3/2}}$$

where μ is the minimum of the absolute value of the linear form

$$\Lambda = A\omega_1{}^2 + B\omega_1\omega_2 + C\omega_2{}^2$$

for integers A, B, C, not all zero, with absolute values at most $c_{27}L$. But

$$|\omega_1^{-2}\Lambda| = |A + B\tau + C\tau^2| \geqslant |\sigma| - \exp(-k^{3/2+17\delta})$$

where

$$\sigma = A + B\alpha + C\alpha^2.$$

Now $\sigma \neq 0$, since if $d \leqslant 2$ by supposition the height of α is exactly $H > c_{27}L$. Thus $|\sigma| > (LH)^{-c_{28}}$, which implies $\mu > H^{-c_{29}}$, and finally

$$|p(\lambda_1,\lambda_2)| \leqslant H^{c_{30}L} \; e^{-k^{3/2}} < 1$$

for all λ_1,λ_2. Since the $p(\lambda_1,\lambda_2)$ are rational integers not all zero this contradiction completes the proof of Theorem 1.

CHAPTER TWO

2.1 Introduction

In this chapter we shall use the result of Theorem I to prove that the dimension of the vector space V defined in the Introduction is six when $\wp$ has no complex multiplication. Thus we shall assume throughout this chapter that $\wp$ has no complex multiplication, and we shall establish the following theorem.

Theorem II

The six numbers 1, ω_1, ω_2, η_1, η_2 and $2\pi i$ are linearly independent over the field $\mathbb{A}$ of algebraic numbers.

2.2 Preliminary Lemmas

We first recall the Wronskian determinant of L+1 meromorphic functions $f^{(0)}(z), \ldots ,f^{(L)}(z)$ given by $\det((d/dz)^{\lambda} f^{(\mu)}(z))$ for $0 \leqslant \lambda,\mu \leqslant L$. Its identical vanishing is a necessary and sufficient condition for the functions to be linearly dependent over the field of complex numbers.

Lemma 2.1

Let $\sigma_0, \ldots ,\sigma_L$ be distinct complex numbers, and let

$$f(z) = \sum_{\lambda=0}^{L} F(\lambda,z)e^{\sigma_\lambda z}$$

for meromorphic functions $F(\lambda,z)$. Then the Wronskian $W(z)$ of the functions

$$f(\mu,z) = \sum_{\lambda=0}^{L} \sigma_\lambda^{\mu} F(\lambda,z)e^{\sigma_\lambda z} \qquad (0 \leqslant \mu \leqslant L)$$

is given by

$$W(z) = \Delta e^{\sigma z} \det_{0 \leqslant \lambda,\mu \leqslant L} F(\lambda,\mu,z)$$

where

$$\Delta = \prod_{\lambda > \mu} (\sigma_\lambda - \sigma_\mu) , \quad \sigma = \sum_{\lambda=0}^{L} \sigma_\lambda ,$$

and the functions $F(\lambda,\mu,z)$ are defined by

$$F(\lambda,0,z) = F(\lambda,z) \qquad (0 \leqslant \lambda \leqslant L)$$

and the recurrence relation

$$F(\lambda,\mu+1,z) = (d/dz)F(\lambda,\mu,z) + \sigma_\lambda F(\lambda,\mu,z) \quad (0 \leqslant \mu < L).$$

Proof

It is easily verified by induction on μ that

$$(d/dz)^{\mu} f(\nu,z) = \sum_{\lambda=0}^{L} \sigma_\lambda^{\nu} F(\lambda,\mu,z)e^{\sigma_\lambda z} ,$$

whence

$$W(z) = \det_{\mu,\lambda} F(\lambda,\mu,z) \det_{\lambda,\nu} (\sigma_\lambda^{\nu} e^{\sigma_\lambda z}).$$

The second factor on the right is essentially a Vandermonde determinant, and the lemma follows.

Lemma 2.2

There is a constant c_1 depending only on ω_1 and ω_2 with the following property. For all $X > c_1$ there are integers A, B, C with

$$\max(|A|,|B|,|C|) = H$$

such that

(i) $e^{-(\log X)^4} < |A\omega_1^2 + B\omega_1\omega_2 + C\omega_2^2| < X^{-1}$,

(ii) $e^{(\log X)^{1/4}} < H < c_2X^2$,

(iii) $|C| > c_3H$.

Proof

This is another application of the Box Principle resembling the proof of Lemma 1.4. When D, E, F are non-negative integers not exceeding $T > c$ (where as before c is sufficiently large) the points $D\omega_1^2 + E\omega_1\omega_2 + F\omega_2^2$ lie in a square centred at the origin with side at most c_4T. This square may be divided into at most $c_5T^2X^2$ smaller disjoint squares of side $\frac{1}{2}X^{-1}$; hence if

$$\tfrac{1}{2}T^3 < c_5T^2X^2 < T^3$$

at least one of these squares contains at least two such points and so their difference

$$\delta = A\omega_1^2 + B\omega_1\omega_2 + C\omega_2^2$$

has absolute value at most X^{-1}. Also

$$H = \max(|A|, |B|, |C|) \leq 2T < c_2X^2,$$

and this proves the right sides of (i) and (ii). For the left sides we denote by α, α' the roots of the equation $A + Bx + Cx^2 = 0$ so that

$$\delta = C\omega_1^2(\tau - \alpha)(\tau - \alpha').$$

Then from Theorem I with $\varepsilon = \frac{1}{2}$ it follows that

$$|\delta| > c_6e^{-2(\log H)^{3/2}} > e^{-(\log X)^4}$$

provided that $X > c$. Also since $|\delta| \leq X^{-1}$ we may assume without loss of generality that

$$|\tau - \alpha| < c_7X^{-1/2}$$

and now Theorem I with $\varepsilon = \frac{1}{2}$ shows that $H > e^{(\log X)^{1/4}}$. It remains to prove (iii). But since τ is not real we have

$$|A\omega_1^2 + B\omega_1\omega_2| \geqslant c_8 \max(|A|,|B|)$$

and the truth of the assertion is now apparent.

For the proof of Theorem II we shall require a lemma concerning the values taken by polynomials at points which are well-distributed in a certain sense. Lemma 1.3 provides such a result for polynomials in one variable, and we now present a more specialized version for three variables. In $\mathbb{C}^3$ we define the absolute value of $\underline{z} = (z_1,z_2,z_3)$ by

$$|\underline{z}|^2 = |z_1|^2 + |z_2|^2 + |z_3|^2$$

and we call $\underline{z}$ a real point if z_1, z_2, z_3 are real. We denote by $\mathcal{B}$ the unit ball $|\underline{z}| \leqslant 1$, and we let $\mathcal{B}_R$ be the set of real points of $\mathcal{B}$.

Lemma 2.3

For an integer $L \geqslant 1$ let $\mathcal{S}$ be a subset of $\mathbb{C}^3$ containing a point within $(21L^2)^{-1}$ of every point of $\mathcal{B}_R$, and for complex numbers $p(\lambda_1,\lambda_2,\lambda_3)$ $(0 \leqslant \lambda_1,\lambda_2,\lambda_3 \leqslant L)$ write

$$\phi(\underline{z}) = \sum_{\lambda_1=0}^{L} \sum_{\lambda_2=0}^{L} \sum_{\lambda_3=0}^{L} p(\lambda_1,\lambda_2,\lambda_3) z_1^{\lambda_1} z_2^{\lambda_2} z_3^{\lambda_3}.$$

Then for all λ_1, λ_2, λ_3 we have

$$|p(\lambda_1,\lambda_2,\lambda_3)| \leqslant (c_{10}L)^{c_{11}L} \sup_{\underline{z} \text{ in } \mathcal{S}} |\phi(\underline{z})|.$$

Proof

We first note the result of Markov [23] that if $f(x)$ is a polynomial of degree at most L then

$$\sup_{-1 \leqslant x \leqslant 1} |f'(x)| \leqslant L^2 \sup_{-1 \leqslant x \leqslant 1} |f(x)|. \tag{17}$$

The result is usually obtained when $f(x)$ has real coefficients, but it is easy to check by inspection that most proofs remain valid when $f(x)$ is in $\mathbb{C}[x]$.

For brevity write $\delta = (21L^2)^{-1}$ and denote by $\mathcal{K}$ the Cartesian product of the sets

$$|\mathrm{Re}(z_i)| \leqslant 1 , \quad |\mathrm{Im}(z_i)| \leqslant \delta \qquad (i = 1,2,3).$$

Also let $\mathcal{K}'$ denote the Cartesian product of the smaller sets

$$|\mathrm{Re}(z_i)| \leqslant 1-\delta , \quad \mathrm{Im}(z_i) = 0 \qquad (i = 1,2,3).$$

We write

$$M(\mathcal{K}) = \sup_{\underline{z} \text{ in } \mathcal{K}} |\phi(\underline{z})| , \quad M(\mathcal{S}) = \sup_{\underline{z} \text{ in } \mathcal{S}} |\phi(\underline{z})|,$$

and we start by showing that $M(\mathcal{K}) \leqslant 2M(\mathcal{S})$.

Let $\underline{\zeta} = (\zeta_1,\zeta_2,\zeta_3)$ be an arbitrary point of $\mathcal{K}$. We construct a point $\underline{\sigma}$ of $\mathcal{S} \cap \mathcal{K}$ close to $\underline{\zeta}$ as follows. Let $\underline{\zeta}'$ be the point of $\mathcal{K}'$ nearest to $\underline{\zeta}$, so that

$$|\underline{\zeta} - \underline{\zeta}'| \leqslant \delta\sqrt{6}.$$

Let $\underline{\sigma}$ be the point of $\mathcal{S}$ nearest $\underline{\zeta}'$, whence $|\underline{\zeta}' - \underline{\sigma}| \leqslant \delta$ by supposition and clearly $\underline{\sigma}$ lies in $\mathcal{K}$. Furthermore

$$|\underline{\zeta} - \underline{\sigma}| \leqslant (1 + \sqrt{6})\delta. \tag{18}$$

We now use the identity

$$\phi(\underline{\zeta}) - \phi(\underline{\sigma}) = \int_{L_1} \phi_1(z_1,\zeta_2,\zeta_3)dz_1 + \int_{L_2} \phi_2(\sigma_1,z_2,\zeta_3)dz_2 + \int_{L_3} \phi_3(\sigma_1,\sigma_2,z_3)dz_3 \tag{19}$$

where $\phi_i = \partial\phi/\partial z_i$ and L_i is a straight line joining σ_i to ζ_i $(i = 1,2,3)$. If $\underline{z}$ lies in $\mathcal{K}$, we have

$$|\phi_i(\underline{z})| \leqslant L^2 M(\mathcal{K}) \qquad (i = 1,2,3) \tag{20}$$

since, e.g., for $i = 1$ we can write $z_1 = \xi_1 + i\eta_1$ with ξ_1, η_1 real and then (20) follows from (17) with

$$f(x) = \phi(x+i\eta_1,z_2,z_3).$$

Thus from (18) and (19) we find that

$$|\phi(\underline{\zeta}) - \phi(\underline{\sigma})| \leqslant 3(1+\sqrt{6})\delta L^2 M(\mathcal{K}) \leqslant \tfrac{1}{2}M(\mathcal{K})$$

and so

$$|\phi(\underline{\zeta})| \leqslant M(\mathcal{S}) + \tfrac{1}{2}M(\mathcal{K}).$$

Because $\underline{\zeta}$ is an arbitrary point of $\mathcal{K}$, this shows that $M(\mathcal{K}) \leqslant 2M(\mathcal{S})$ as asserted.

Finally since $\mathcal{K}$ is a Cartesian product we can apply Lemma 1.3 to each of the variables z_1, z_2, z_3 in turn (with, e.g., $\sigma_\lambda = \lambda/L$ for $0 \leqslant \lambda \leqslant L$) to deduce the estimate of the present lemma.

2.3 The Main Lemma

The purpose of this section is to prove the following lemma, which is a weaker version of Lemma 1.5 extending to three functions. Let $\overline{\beta}$ denote the complex conjugate of β.

Lemma 2.4

Let β_1, β_2 be complex numbers such that

$$\overline{\beta_1}\omega_1 + \overline{\beta_2}\omega_2 \neq 0,$$

and write

$$f(z) = \beta_1(\zeta(\omega_1 z) - \eta_1 z) + \beta_2(\zeta(\omega_2 z) - \eta_2 z).$$

For an integer $L \geqslant 1$ and complex numbers $p(\lambda_0,\lambda_1,\lambda_2)$ $(0 \leqslant \lambda_0,\lambda_1,\lambda_2 \leqslant L)$ let

$$\phi(z) = \sum_{\lambda_0=0}^{L}\sum_{\lambda_1=0}^{L}\sum_{\lambda_2=0}^{L} p(\lambda_0,\lambda_1,\lambda_2)(f(z))^{\lambda_0}(\wp(\omega_1 z))^{\lambda_1}(\wp(\omega_2 z))^{\lambda_2}$$

and for $r \geqslant 0$ let $M(r)$ denote the maximum modulus of $\phi(z)$ at points of $\mathcal{E}$ with $|z| \leqslant r$. Then for all λ_0, λ_1, λ_2 we have

$$|p(\lambda_0,\lambda_1,\lambda_2)| \leqslant (c_{12}L)^{c_{13}L} M(c_{14}\exp((\log L)^{36}))$$

where c_{12}, c_{13} and c_{14} depend only on β_1, β_2, ω_1 and ω_2.

Proof

Without loss of generality we may suppose that $L > c$ for a sufficiently large constant c and also $\beta_2 \neq 0$. We use Lemma 2.2 to find integers A, B, C, not all zero, and integers D, E, F, not all zero with

$$\max(|A|,|B|,|C|) \leqslant L^{21},$$

$$\max(|D|,|E|,|F|) \leqslant \exp((\log L)^{34})$$

such that if we write

$$\delta = A\omega_1{}^2 + B\omega_1\omega_2 + C\omega_2{}^2, \quad \varepsilon = D\omega_1{}^2 + E\omega_1\omega_2 + F\omega_2{}^2$$

then

$$|\delta| \leqslant L^{-10}, \quad |\varepsilon| \leqslant \exp(-(\log L)^{33}).$$

We set $\theta = \delta/|\delta|$ and for integers r_0, r_1, r_2 with absolute values at most L^3 we write

$$z(r_0,r_1,r_2) = \tfrac{1}{4} + \theta r_1/\omega_1\omega_2 L^7 + (mC + r_0 F)\tau$$

where m is the integer nearest to $(r_2 - r_1)/|\delta|L^7$. Since $|m\delta| < 2L^{-4}$ the equations

$$\omega_1 z(r_0,r_1,r_2) = \tfrac{1}{4}\omega_1 + \theta r_1/\omega_2 L^7 + (mC + r_0 F)\omega_2 \tag{21}$$

and

$$\omega_2 z(r_0,r_1,r_2) = \tfrac{1}{4}\omega_2 + \theta r_1/\omega_1 L^7 + m\delta/\omega_1 + r_0\varepsilon/\omega_1 - (mA + r_0 D)\omega_1 - (mB + r_0 E)\omega_2 \tag{22}$$

show that $z(r_0,r_1,r_2)$ lies in $\mathcal{E}$. Furthermore from (i) of Lemma 2.2 we have

$$|\delta| > \exp(-(\log L)^5),$$

whence

$$|m| < \exp((\log L)^5)$$

and so

$$|z(r_0,r_1,r_2)| < \exp((\log L)^{35}).$$

It follows that for all r_0, r_1, r_2

$$|\phi(z(r_0,r_1,r_2))| \leqslant M = M(\exp((\log L)^{35})),$$

and we shall base our arguments on these inequalities.

For an integer r with absolute value at most L^3 we write

$$z(r) = \tfrac{1}{4} + \theta r/\omega_1\omega_2 L^7$$

and

$$x_i(r) = \wp(\omega_i z(r)) \qquad (i = 1,2)$$

and we first verify the inequalities

$$|\wp(\omega_i z(r_0,r_1,r_2)) - x_i(r_i)| < L^{-8} \qquad (i = 1,2). \tag{23}$$

In fact for $i = 1$ the left side of this vanishes, by (21). Also from the definition of m we have

$$|\theta r_1/\omega_1 L^7 + m\delta/\omega_1 - \theta r_2/\omega_1 L^7|$$
$$= |\delta\omega_1^{-1}|\,|m - (r_2 - r_1)/|\delta|L^7| < c_{15}|\delta|$$

so that from (22) and Lemma 1.1 we deduce that the left side of (23) for $i = 2$ is at most

$$c_{16}(|\delta| + L^3|\varepsilon|) < L^{-8}.$$

Next we examine the behaviour of $f(z)$ at the points $z(r_0,r_1,r_2)$, and for brevity we write

$$g_i(z) = \zeta(\omega_i z) - \eta_i z, \qquad (i = 1,2)$$

whence

$$f(z) = \beta_1 g_1(z) + \beta_2 g_2(z).$$

Then on using the Legendre relation

$$\omega_2\eta_1 - \omega_1\eta_2 = 2\pi i$$

we see from (21) that

$$g_1(z(r_0,r_1,r_2)) + 2\pi i r_0 F/\omega_1 = g_1(z(r_1)) - 2\pi i m C/\omega_1. \tag{24}$$

Similarly if

$$z' = z(r_1) + (m\delta + r_0\varepsilon)/\omega_1\omega_2$$

we find from (22) that

$$g_2(z(r_0,r_1,r_2)) + 2\pi i r_0 D/\omega_2 = g_2(z') - 2\pi i m A/\omega_2. \qquad (25)$$

Since the points z', $z(r_1)$ lie in $\mathcal{D}$, the corresponding values of g_1 and g_2 are bounded, and hence from (24) and (25)

$$|f(z(r_0,r_1,r_2)) - r_0\Delta| < c_{17}(1 + |mA| + |mC|) < \exp((\log L)^6) \qquad (26)$$

where

$$\Delta = -2\pi i(\beta_1 F/\omega_1 + \beta_2 D/\omega_2).$$

It is now necessary to derive a lower bound for $|\Delta|$ that justifies treating the right side of (26) as an error term. For brevity write $\beta = -\beta_1/\beta_2$. Then

$$\beta\tau = D/F + \omega_2\Delta/2\pi i\beta_2 F$$

and

$$\beta + \tau = -E/F + \varepsilon/\omega_1\omega_2 F + \omega_1\Delta/2\pi i\beta_2 F$$

and these imply

$$|\mathrm{Im}(\beta\tau)| \leqslant c_{18}|\Delta|/|F|, \quad |\mathrm{Im}(\beta+\tau)| \leqslant c_{19}(1 + |\Delta|)/|F|. \qquad (27)$$

By hypothesis β and τ are not complex conjugates, and since τ is not real, this means that not both of the left sides of (27) can vanish. Therefore

$$|\Delta| \geqslant c_{20}|F| - 1$$

and a lower bound for $|F|$ is provided by (ii) and (iii) of Lemma 2.2; we find that

$$|F| > \exp((\log L)^8)$$

whence

$$|\Delta| > \exp((\log L)^7). \qquad (28)$$

The inequalities (23) and (26) express the fact that the values of the functions $f(z)$, $\wp(\omega_1 z)$ and $\wp(\omega_2 z)$ at the points $z(r_0,r_1,r_2)$ are in a sense distributed independently as r_0, r_1 and r_2 vary. By a simple change of variables we can achieve the situation of Lemma 2.3. Let

$$b_0 = L^3\Delta\ ,\quad a_i = \wp(\tfrac{1}{4}\omega_i)\ ,\quad b_i = \theta\omega_i\wp'(\tfrac{1}{4}\omega_i)/\omega_1\omega_2 L^4 \qquad (i = 1,2)$$

and define the polynomial

$$Q(x_0,x_1,x_2) = \sum_{\lambda_0=0}^{L}\sum_{\lambda_1=0}^{L}\sum_{\lambda_2=0}^{L} p(\lambda_0,\lambda_1,\lambda_2)(b_0x_0)^{\lambda_0}(a_1+b_1x_1)^{\lambda_1}(a_2+b_2x_2)^{\lambda_2}.$$

From the power series expansion of $\wp(\omega_i z)$ about $z = \frac{1}{4}$ we have

$$|x_i(r_i) - a_i - b_ir_i/L^3| < c_{21}L^{-8} \qquad (i = 1,2)$$

and hence the point (ξ_0,ξ_1,ξ_2) of $\mathbb{C}^3$ defined by

$$b_0\xi_0 = f(z(r_0,r_1,r_2)),$$

$$a_i + b_i\xi_i = \wp(\omega_i z(r_0,r_1,r_2)) \qquad (i = 1,2)$$

satisfies, from (23), (26) and (28)

$$|\xi_i - r_i/L^3| < c_{22}L^{-4} \qquad (i = 0,1,2).$$

Now if (η_1,η_2,η_3) is any point of $\mathcal{B}_R$ there are integers r_0, r_1, r_2 with absolute values at most L^3 such that

$$|\eta_i - r_i/L^3| \leq 1/2L^3 \qquad (i = 0,1,2).$$

It follows that within $2L^{-3}$ of each point of $\mathcal{B}_R$ there is a point (ξ_0,ξ_1,ξ_2) with

$$|Q(\xi_0,\xi_1,\xi_2)| = |\phi(z(r_0,r_1,r_2))| \leq M,$$

and therefore if $q(\lambda_0,\lambda_1,\lambda_2)$ denotes the coefficient of $x_0^{\lambda_0}x_1^{\lambda_1}x_2^{\lambda_2}$ in $Q(x_0,x_1,x_2)$ we see from Lemma 2.3 that

$$|q(\lambda_0,\lambda_1,\lambda_2)| \leq L^{c_{23}L}M$$

for all λ_0, λ_1, λ_2.

Finally since

$$p(\lambda_0,\lambda_1,\lambda_2) = \sum_{\mu_1=\lambda_1}^{L}\sum_{\mu_2=\lambda_2}^{L}\binom{\mu_1}{\lambda_1}\binom{\mu_2}{\lambda_2}b_0^{-\lambda_0}(-a_1)^{\mu_1-\lambda_1}b_1^{-\mu_1}(-a_2)^{\mu_2-\lambda_2} b_2^{-\mu_2}q(\lambda_0,\mu_1,\mu_2)$$

we conclude that

$$|p(\lambda_0,\lambda_1,\lambda_2)| \leq L^{c_{24}L} M,$$

and this completes the proof of the lemma.

2.4 The Auxiliary Function

We commence the proof of Theorem II by constructing the auxiliary function. We have seen that in [12] Coates proved that a non-zero linear combination of the numbers ω_1, ω_2, η_1, η_2 and $2\pi i$ is transcendental. We shall show that the equation

$$\alpha_1\omega_1 + \alpha_2\omega_2 + \beta_1\eta_1 + \beta_2\eta_2 + \gamma.2\pi i = 0 \tag{29}$$

is impossible if at least one of the algebraic numbers α_1, α_2, β_1, β_2, γ is non-zero. Accordingly we assume that (29) holds with algebraic numbers α_1, α_2, β_1, β_2 and γ, and from the result of Coates [13] establishing the impossibility of (29) with $\beta_1 = \beta_2 = 0$ it is clear that we may suppose $\beta_2 \neq 0$. Further, as in [12], we suppose that $\frac{1}{4}g_2$, $\frac{1}{4}g_3$ and the coefficients in (29) are algebraic integers, and as usual this involves no loss of generality. We shall eventually derive a contradiction, and this will prove Theorem II.

For a large integer k we set

$$L = [k^{4/5}], \quad h = [k^{1/10}]$$

and we write

$$\Phi(z_1,z_2,z_3) = \sum_{\lambda_0=0}^{L}\sum_{\lambda_1=0}^{L}\sum_{\lambda_2=0}^{L}\sum_{\lambda_3=0}^{L} p(\lambda_0,\lambda_1,\lambda_2,\lambda_3)(f(z_1,z_2,z_3))^{\lambda_0} (\wp(\omega_1 z_1))^{\lambda_1}(\wp(\omega_2 z_2))^{\lambda_2} e^{2\pi i\lambda_3 z_3},$$

where

$$f(z_1,z_2,z_3) = \alpha_1\omega_1 z_1 + \alpha_2\omega_2 z_2 + \beta_1\zeta(\omega_1 z_1) + \beta_2\zeta(\omega_2 z_2) + 2\pi i\gamma z_3$$

and the coefficients $p(\lambda_0,\lambda_1,\lambda_2,\lambda_3)$ are yet to be determined. Formally this is the auxiliary function used by Coates [12], but his assumption is that the right side of (29) is an algebraic number $\alpha_0 \neq 0$. However, the vanishing of α_0 does not invalidate his arguments except in the concluding paragraphs. Thus denoting by $c, c_1, \ldots$ positive constants depending only on α_1, α_2, β_1, β_2, γ, ω_1 and ω_2 we see that for $k > c$ the following lemmas hold.

Lemma 2.5

There exist rational integers $p(\lambda_0,\lambda_1,\lambda_2,\lambda_3)$, not all zero, with absolute values at most k^{10k}, such that

$$\Phi_{m_1,m_2,m_3}(s+\tfrac{1}{2},s+\tfrac{1}{2},s+\tfrac{1}{2}) = 0$$

for all integers s with $1 \leqslant s \leqslant h$ and all non-negative integers m_1, m_2, m_3 with

$$m_1 + m_2 + m_3 \leqslant k.$$

Proof

See Lemma 7 of [12].

Lemma 2.6

Suppose that $Z \geqslant 6$ and let

$$\phi(z_1,z_2,z_3) = \Phi(z_1,z_2,z_3) \prod_{i=1}^{2} \prod_{\Omega_i} (\omega_i z_i - \Omega_i)^{3L}$$

where Ω_i runs over all poles of $\wp(z)$ with $|\Omega_i| \leqslant |\omega_i| Z$. Then $\phi(z_1,z_2,z_3)$ is regular in the disc $|z_i| \leqslant Z$ $(i = 1,2,3)$, and for any z with $|z| \leqslant \tfrac{1}{2}Z$ and for any non-negative integers m_1, m_2, m_3 with $m_1 + m_2 + m_3 \leqslant k$ we have

$$|\phi_{m_1,m_2,m_3}(z,z,z)| < k^{12k} Z^{c_1 L Z^2}.$$

Proof

See Lemma 8 of [12].

Lemma 2.7

Let Q, S, Z be numbers with $1 < Q < S < Z-1$, and let m_1, m_2, m_3 be non-negative integers with $m_1 + m_2 + m_3 \leqslant k$. Suppose that q, r, s are integers with q even, $(r,q) = 1$, and

$$1 \leqslant q \leqslant Q\ ,\quad 1 \leqslant s \leqslant S\ ,\quad 1 \leqslant r < q$$

such that

$$\phi_{\mu_1,\mu_2,\mu_3}(s+r/q, s+r/q, s+r/q) = 0 \tag{30}$$

for all non-negative integers μ_1, μ_2, μ_3 with

$$\mu_1 + \mu_2 + \mu_3 < m_1 + m_2 + m_3.$$

Then either (30) holds with $\mu_i = m_i$ $(i = 1,2,3)$, or else

$$|\phi_{m_1,m_2,m_3}(s+r/q, s+r/q, s+r/q)| > (kS)^{-c_2 k Q^4}\ .$$

Proof

See Lemma 9 of [12] .

The next lemma replaces Lemma 10 of [12] and gives a rather larger range of zeros for Φ without substantially lowering their multiplicities.

Lemma 2.8

Let J be an integer satisfying $0 \leqslant J \leqslant (\log k)^{50}$. Then

$$\Phi_{m_1,m_2,m_3}(s+r/q, s+r/q, s+r/q) = 0$$

for all integers s, r, q with q even, $(r,q) = 1$,

$$1 \leqslant q \leqslant 2h^{J/8}\ ,\quad 1 \leqslant s \leqslant h^{1+J/4}\ ,\quad 1 \leqslant r < q$$

and all non-negative integers m_1, m_2, m_3 with

$$m_1 + m_2 + m_3 \leqslant k - Jk^{39/40}\ .$$

Proof

The lemma is valid for $J = 0$ from Lemma 2.5. We suppose that I is an integer with $0 \leqslant I \leqslant (\log k)^{50} - 1$, and we assume that the lemma holds for $0 \leqslant J < I$. We proceed to deduce its validity for $J = I+1$.

We define

$$\kappa = [k^{39/40}] \;,\quad Q_J = 2h^{J/8} \;,\quad S_J = h^{1+J/4} \;,\quad T_J = [k - J\kappa] \;;$$

then if the lemma is false for $J = I+1$, integers s', r', q' exist with q' even, $(r',q') = 1$,

$$1 \leqslant q' \leqslant Q_{I+1} \;,\quad 1 \leqslant s' \leqslant S_{I+1} \;,\quad 1 \leqslant r' < q'$$

such that

$$\Phi_{m_1',m_2',m_3'}(s'+r'/q',s'+r'/q',s'+r'/q') \neq 0 \tag{31}$$

where m_1', m_2', m_3' are non-negative integers with $m_1' + m_2' + m_3' \leqslant T_{I+1}$. Further we assume that m_1', m_2', m_3' are chosen minimally so that

$$\Phi_{\mu_1,\mu_2,\mu_3}(s'+r'/q',s'+r'/q',s'+r'/q') = 0$$

for all non-negative integers μ_1, μ_2, μ_3 with

$$\mu_1 + \mu_2 + \mu_3 < m_1' + m_2' + m_3' .$$

The contradiction that we shall eventually derive from these assumptions will establish the lemma.

Let $Z = 10S_{I+1}$ and let $\phi(z_1,z_2,z_3)$ be the function defined in Lemma 2.6 for this value of Z. We write

$$\psi(z) = \phi_{m_1',m_2',m_3'}(z,z,z).$$

Then from our induction hypothesis we see that for all integers s, r, q with q even, $(r,q) = 1$

$$1 \leqslant q \leqslant Q_I \;,\quad 1 \leqslant s \leqslant S_I \;,\quad 1 \leqslant r < q$$

and each integer m with $0 \leqslant m \leqslant \kappa$, we have

$$\psi_m(s + r/q) = 0, \tag{32}$$

for the left side is given by

$$\sum_{\mu_1=0}^{m}\sum_{\mu_2=0}^{m}\sum_{\substack{\mu_3=0\\ \mu_1+\mu_2+\mu_3=m}}^{m} m!(\mu_1!\mu_2!\mu_3!)^{-1}\,\phi_{m_1'+\mu_1,m_2'+\mu_2,m_3'+\mu_3}(s+r/q,s+r/q,s+r/q)$$

and the partial derivatives here vanish since they may be expanded in partial derivatives of Φ of order at most

$$\max_{\mu_0,\mu_1,\mu_3} (m_1' + \mu_1 + m_2' + \mu_2 + m_3' + \mu_3) \leqslant T_{I+1} + \kappa \leqslant T_I .$$

For brevity we write

$$F(z) = \prod_{\substack{q=1\\ q \text{ even}}}^{Q_I} \prod_{s=1}^{S_I} \prod_{\substack{r=1\\ (r,q)=1}}^{q} (z - s - r/q)^{\kappa} .$$

Then by (32) the function $\psi(z)/F(z)$ is regular in the disc $|z| \leqslant 5S_{I+1}$. Hence, denoting by θ and Θ the upper bound of $|\psi(z)|$ and the lower bound of $|F(z)|$ on the circle $|z| = 5S_{I+1}$, we conclude from the maximum modulus principle that

$$|\psi(s' + r'/q')| \leqslant |F(s' + r'/q')|\theta/\Theta .$$

Now Lemma 2.6 provides the upper bound

$$\theta \leqslant k^{12k} (10S_{I+1})^{c_3 L S_{I+1}^2} .$$

Further, for any z with $|z| = 5S_{I+1}$ we have

$$|z - s - r/q| \geqslant 2|(s' + r'/q') - (s + r/q)|$$

for each factor of $F(z)$; the number of sets s, r, q occuring here is at least $c_4 S_I Q_I^2$, and so

$$\Theta > 2^{c_5 \kappa S_I Q_I^2} |F(s' + r'/q')| .$$

Hence on noting the estimates

$$\log S_{I+1} < (\log k)^{52} , \quad L S_{I+1}^2 < c_6 h^{\frac{41}{2}+\frac{1}{2}I} , \quad \kappa S_I Q_I^2 > c_7 h^{\frac{43}{4}+\frac{1}{2}I}$$

we deduce that

$$|\psi(s' + r'/q')| < 2^{-c_8 \kappa S_I Q_I^2} .$$

On the other hand, the hypotheses of Lemma 2.7 are satisfied with $Q = Q_{I+1}$, $S = S_{I+1}$ and primed letters throughout; since $\psi(s' + r'/q')$ is the left side of (31) multiplied by a non-zero factor we have the lower bound

$$|\psi(s' + r'/q')| > (kS_{I+1})^{-c_9 k Q_{I+1}^4} .$$

But

$$kQ_{I+1}^{4} < c_{10}\, h^{\frac{3L}{2}+\frac{1}{2}I}$$

and therefore this is inconsistent with the upper bound. This contradiction proves the lemma.

For the next lemma we write $K = \exp((\log k)^{50})$.

Lemma 2.9

For all points z of $\mathcal{E}$ with absolute value at most K and all integers m with $0 \leqslant m \leqslant L$ we have

$$|\Phi_{0,0,m}(z,z,z)| < e^{-K^4}.$$

Proof

Let ζ be any point of $\mathcal{E}$ with $|\zeta| \leqslant K$. Now the least integer I such that $S_I \geqslant K^2$ certainly satisfies

$$I + 1 \leqslant (\log k)^{50}.$$

Therefore if we define the function $\phi(z_1,z_2,z_3)$ with $Z = 10\, S_{I+1}$ and if for any integer μ with $0 \leqslant \mu \leqslant L$ we write

$$\psi(z) = \phi_{0,0,\mu}(z,z,z)$$

we can deduce the inequalities (32) from Lemma 2.8. As in that lemma, the maximum modulus principle gives the estimate

$$|\psi(\zeta)| < 2^{-c_{11} k S_I Q_I^2} < e^{-K^4}.$$

Finally

$$\psi(\zeta) = P\Phi_{0,0,\mu}(\zeta,\zeta,\zeta)$$

where

$$P = \prod_{i=1}^{2} \prod_{\Omega_i} (\omega_i\zeta - \Omega_i)^{3L}$$

and the product is over all poles Ω_i with $|\Omega_i| \leqslant |\omega_i|Z$ $(i = 1,2)$. Clearly there are at most c_{12} periods Ω_i such that $|\omega_i\zeta - \Omega_i| \leqslant 2$, and since ζ lies in $\mathcal{E}$ we have

$|\omega_i\zeta - \Omega_i| > c_{13}$ for these periods. Hence $|P| > 1$ and the lemma follows.

2.5 Proof of Theorem II

We define the functions

$$F(\lambda_3) = F(\lambda_3, z_1, z_2, z_3) = \sum_{\lambda_0=0}^{L} \sum_{\lambda_1=0}^{L} \sum_{\lambda_2=0}^{L} p(\lambda_0, \lambda_1, \lambda_2, \lambda_3)(f(z_1, z_2, z_3))^{\lambda_0} (\wp(\omega_1 z_1))^{\lambda_1} (\wp(\omega_2 z_2))^{\lambda_2}$$

so that

$$\Phi(z_1, z_2, z_3) = \sum_{\lambda_3=0}^{L} F(\lambda_3) e^{2\pi i \lambda_3 z_3} .$$

We assume that not all the functions $F(\lambda_3)$ vanish identically and we shall deduce a contradiction. Suppose $r_0, \dots, r_M$ are the distinct integers between 0 and L such that $F(r_\mu)$ is not identically zero; thus $0 \leq M \leq L$ and

$$\Phi(z_1, z_2, z_3) = \sum_{\mu=0}^{M} F(r_\mu) e^{2\pi i r_\mu z_3} .$$

We are going to use Lemma 2.1 to construct a new function Ψ in which the exponential terms are absent; to this end we define the functions

$$G(\mu, \nu) = G(\mu, \nu, z_1, z_2, z_3) \qquad (0 \leq \mu, \nu \leq M)$$

by

$$G(\mu, 0) = F(r_\mu) \qquad (0 \leq \mu \leq M)$$

and

$$G(\mu, \nu+1) = (\partial/\partial z_3)G(\mu, \nu) + 2\pi i r_\mu G(\mu, \nu) \quad (0 \leq \nu < M).$$

Lemma 2.10

Let

$$\Psi(z_1, z_2, z_3) = (2\pi i)^{-\frac{1}{2}M(M+1)} \det_{\mu,\nu} G(\mu, \nu) .$$

Then for all z in $\mathcal{E}$ with absolute value at most K we have

$$|\Psi(z, z, z)| < e^{-K^2} .$$

Proof

From Lemma 2.1 we see that

$$\Psi(z_1,z_2,z_3) = (2\pi i)^{-\frac{1}{2}M(M+1)} R^{-1} e^{-\sigma z_3} W(z_1,z_2,z_3) \qquad (33)$$

where

$$R = \prod_{\mu>\nu} (r_\mu - r_\nu) \ , \quad \sigma = 2\pi i \sum_{\mu=0}^{M} r_\mu$$

and $W(z_1,z_2,z_3)$ is the Wronskian of the functions $\Phi(0), \ldots ,\Phi(M)$ with respect to z_3, where

$$\Phi(\nu) = \Phi(\nu,z_1,z_2,z_3) = \sum_{\mu=0}^{M} (2\pi i r_\mu)^\nu F(r_\mu) e^{2\pi i r_\mu z_3} .$$

We proceed to establish an upper bound for $|W(z,z,z)|$ when z satisfies the conditions of the lemma. The first column of the determinant defining $W(z,z,z)$ consists of the terms $\Phi_{0,0,\mu}(z,z,z)$ and from Lemma 2.9 these are at most e^{-K^4} in absolute value. To estimate the other terms we observe that if z_3 is a complex number with $|z_3 - z| = 1$

$$|\Phi(\nu,z,z,z_3)| < K^{c_{14}LK^2}$$

and we have the integral (recall the integrand is regular in z_3)

$$(\partial/\partial z_3)^\mu \Phi(\nu,z,z,z) = \frac{\mu!}{2\pi i} \int \frac{\Phi(\nu,z,z,z_3)}{(z_3 - z)^{\mu+1}} dz_3$$

taken over the positively described unit circle centred at z. Thus the other entries in the Wronskian determinant are at most $K^{c_{15}LK^2}$ in absolute value, and it follows that

$$|W(z,z,z)| \leqslant K^{c_{16}L^2K^2} \ e^{-K^4} < e^{-K^3} .$$

Finally since

$$|R| \geqslant 1 \ , \quad |e^{-\sigma z}| \leqslant e^{8L^2K}$$

we deduce the inequality of the lemma immediately from (33).

It is easy to verify by induction on ν that

$$G(\mu,\nu) = (2\pi i)^{\nu} \sum_{\lambda_0=0}^{L} \sum_{\lambda_1=0}^{L} \sum_{\lambda_2=0}^{L} \sum_{m=0}^{\nu} q(\lambda_0,\lambda_1,\lambda_2,m)(f(z_1,z_2,z_3))^{\lambda_0-m} (\wp(\omega_1 z_1))^{\lambda_1}(\wp(\omega_2 z_2))^{\lambda_2},$$

where

$$q = q(\lambda_0,\lambda_1,\lambda_2,m) = \binom{\nu}{m} r_\mu^{\nu-m} \gamma^m \lambda_0(\lambda_0-1) \ldots (\lambda_0-m+1)p(\lambda_0,\lambda_1,\lambda_2,r_\mu)$$

if $m \leq \lambda_0$ and q is zero otherwise. Thus we see that the coefficients q are integers of the field $\mathbb{Q}(\gamma)$ whose sizes are at most $k^{c_{17}k}$. It follows that the function Ψ may be written in the form

$$\Psi(z_1,z_2,z_3) = \sum_{\nu_0=0}^{N} \sum_{\nu_1=0}^{N} \sum_{\nu_2=0}^{N} r(\nu_0,\nu_1,\nu_2)(f(z_1,z_2,z_3))^{\nu_0}(\wp(\omega_1 z_1))^{\nu_1} (\wp(\omega_2 z_2))^{\nu_2}$$

where $N = L(L+1)$ and the coefficients $r(\nu_0,\nu_1,\nu_2)$ are algebraic integers of $\mathbb{Q}(\gamma)$ with sizes at most $c_{18}{}^{k^2}$. We now observe that in view of (29)

$$f(z,z,z) = \beta_1(\zeta(\omega_1 z) - \eta_1 z) + \beta_2(\zeta(\omega_2 z) - \eta_2 z) \quad (34)$$

and further $\overline{\beta_1}\omega_1 + \overline{\beta_2}\omega_2 \neq 0$ since $\beta_2 \neq 0$ and ω_2/ω_1 is transcendental. Therefore from Lemma 2.4 and Lemma 2.10 it follows that

$$|r(\nu_0,\nu_1,\nu_2)| \leq N^{c_{19}N} e^{-K^3} < e^{-K}$$

for all ν_0, ν_1 and ν_2. If $r(\nu_0,\nu_1,\nu_2) \neq 0$ this would contradict the obvious lower bound $c_{20}{}^{-k^2}$; hence $r(\nu_0,\nu_1,\nu_2) = 0$ for all ν_0, ν_1, ν_2 and $\Psi(z_1,z_2,z_3)$ is identically zero. From (33) the Wronskian $W(z_1,z_2,z_3)$ vanishes identically and so by the basic property of Wronskians there exist meromorphic functions $H(\nu) = H(\nu,z_1,z_2)$ $(0 \leq \nu \leq M)$ independent of z_3 and not all zero, such that

$$\sum_{\nu=0}^{M} H(\nu)\Phi(\nu) = 0.$$

Thus we have

$$\sum_{\mu=0}^{M} E(\mu)F(r_\mu)e^{2\pi i r_\mu z_3} = 0$$

where

$$E(\mu) = E(\mu,z_1,z_2) = \sum_{\nu=0}^{M} (2\pi i r_\mu)^\nu H(\nu) \qquad (35)$$

Since the exponential function is not an algebraic function, this implies that $E(\mu)F(r_\mu) = 0$ for all μ and therefore $E(\mu)$ vanishes identically for all μ. But the determinant of the equations (35) is of Vandermonde type and plainly non-zero; it follows that $H(\nu) = 0$ for all ν. In view of the assumption made at the beginning of this section, this contradiction implies that $F(\lambda_3)$ vanishes identically for all λ_3. But, using (34) again, the function $F(\lambda_3,z,z,z)$ is of the type considered in Lemma 2.4. A rather trivial application of this lemma shows that all the coefficients $p(\lambda_0,\lambda_1,\lambda_2,\lambda_3)$ must vanish, and this contradiction finally establishes Theorem II.

As a postscript we remark that this last use of Lemma 2.4 is a little unnecessary, for it is easy to prove directly the algebraic independence of the functions $f(z_1,z_2,z_3)$, $\wp(\omega_1 z_1)$ and $\wp(\omega_2 z_2)$. (See, e.g., [18], p.111.)

CHAPTER THREE

3.1 Introduction

In this chapter we assume that $\wp(z)$ has complex multiplication over the complex quadratic field $\mathbb{K}$, and our principal object is to prove the following theorem.

Theorem III

The dimension of V over $\mathbb{A}$ is four and the numbers 1, ω_1, η_1 and $2\pi i$ are a basis for V.

3.2 The Upper Bound

We show in this section, among other things, that the dimension of V is at most four. Since the ratio $\tau = \omega_2/\omega_1$ lies in $\mathbb{K}$, there exist coprime integers A,B,C $\neq 0$ such that

$$A + B\tau + C\tau^2 = 0,$$

and the relation $\tau\omega_1 - \omega_2 = 0$ implies that the dimension of V is at most five. To proceed further we prove the following lemma.

Lemma 3.1

There is an algebraic number κ of the field $\mathbb{K}(g_2, g_3)$ such that

$$A\eta_1 - C\tau\eta_2 = \kappa\omega_2. \tag{36}$$

Proof

We define κ by (36) and consider the function

$$f(z) = -A\zeta(Cz) + C\tau\zeta(C\tau z) + C\tau\kappa z. \qquad (37)$$

Then we have

$$f(z + \omega_1) - f(z) = -AC\eta_1 + C^2\tau\eta_2 + C\tau\kappa\omega_1 = 0$$

and, since

$$C\tau\omega_2 = -A\omega_1 - B\omega_2$$

we also have

$$f(z + \omega_2) - f(z) = -AC\eta_2 + C\tau(-A\eta_1 - B\eta_2) + C\tau\kappa\omega_2 = 0.$$

Therefore f(z) is a doubly periodic function with the same periods as $\wp(z)$, and it follows that f(z) is a rational function of $\wp(z)$ and $\wp'(z)$. Let σ be an arbitrary monomorphism of $\mathbb{K}(g_2, g_3, \kappa)$ fixing the field $\mathbb{F} = \mathbb{K}(g_2, g_3)$, and let $f^{\sigma}(z)$ be the function obtained by applying this to the Laurent expansion of f(z) about the origin. Since σ acting in this way leaves $\wp(z)$ and $\wp'(z)$ fixed, we deduce that $f^{\sigma}(z)$ is also a rational function of $\wp(z)$ and $\wp'(z)$. Hence the function

$$f(z) - f^{\sigma}(z) = C\tau(\kappa - \kappa^{\sigma})z$$

is an elliptic function, which is only possible if $\kappa = \kappa^{\sigma}$. Since σ is arbitrary, this proves that κ lies in $\mathbb{F}$.

The preceeding lemma implies that the numbers 1, ω_1, η_1 and $2\pi i$ certainly span V; hence Theorem III will follow from the linear independence of these numbers over $\mathbb{A}$. This will be proved in the next section after we have derived an additional property of the number κ.

Lemma 3.2

The algebraic number κ vanishes if and only if one of the invariants g_2, g_3 vanishes.

Proof

For any complex number z in the upper half plane $\mathrm{Im}(z) > 0$ we write $q = e^{2\pi iz}$ and define the function

$$E_2(z) = 1 - 24 \sum_{n=1}^{\infty} nq^n(1 - q^n)^{-1} = 1 - 24 \sum_{n=1}^{\infty} \sigma(n)q^n$$

where $\sigma(n)$ denotes the sum of the divisors of n. This is a well-known function with modular properties, and it satisfies the functional equation (e.g., [24])

$$E_2(-1/z) = z^2E_2(z) + 6z/\pi i \ . \tag{38}$$

Furthermore we have

$$E_2(\tau) = 3\omega_1\eta_1/\pi^2.$$

From this together with the Legendre relation or its equivalent (38) we obtain the formula

$$\kappa = -\pi^2(B + 2C\tau)\phi(\tau)/3\omega_1{}^2$$

where

$$\phi(z) = E_2(z) - 3/(\pi \mathrm{Im}\, z).$$

Therefore the vanishing of κ implies that $\phi(\tau) = 0$, and we proceed to find all the zeros of $\phi(z)$ in the upper half plane. From (38) it is easily seen that $\phi(z)$ has period 1 and

$$\phi(-1/z) = z^2\phi(z), \tag{39}$$

so that the set of zeros is mapped into itself by the modular group Γ. Hence we may confine our search to a fundamental region $\mathcal{R}$ for Γ, and we take this as the region defined by

$$-\tfrac{1}{2} \leqslant s < \tfrac{1}{2}, \quad |z| \geqslant 1$$

where $z = s + it$ with s, t real.

We start by observing that $z = i$ and $z = \rho = \frac{1}{2}(-1 + \sqrt{-3})$ are zeros of $\phi(z)$. This is clear on putting $z = i$ and $z = 1 + \rho$ in (39), and so $\kappa = 0$ when τ has these values; they correspond to $g_3 = 0$ and $g_2 = 0$ respectively. Next we assert that $E_2(z)$ is real if and only if $s = 0$ or $-\frac{1}{2}$. This is a consequence of the expansion for $s \neq 0, -\frac{1}{2}$

$$\mathrm{Im}(E_2(z)) = -24e^{-2\pi t} \sin 2\pi s(1 + g(s,t))$$

where

$$g(s,t) = \sum_{n=2}^{\infty} \sigma(n) e^{-2\pi(n-1)t} \quad \sin 2\pi ns/\sin 2\pi s$$

together with the inequalities

$$|\sin 2\pi ns/\sin 2\pi s| \leqslant n, \quad \sigma(n) < n^2, \quad (n+2)^3 < 2(n+1)(n+2)(n+3)$$

$$e^{-2\pi t} < \gamma = e^{-\pi\sqrt{3}} < (200)^{-1}.$$

For then

$$|g(s,t)| < 2\gamma \sum_{n=0}^{\infty} (n+1)(n+2)(n+3)\gamma^n = 12\gamma(1-\gamma)^{-4} < \tfrac{1}{5}$$

and so $1 + g(s,t)$ cannot vanish.

Thus a zero of $\phi(z)$ must have $s = 0$ or $-\frac{1}{2}$, and we treat each possibility in turn. If $s = 0$, we obtain

$$E_2(it) = 1 - 24 \sum_{n=1}^{\infty} \sigma(n) e^{-2\pi nt} = 3/\pi t,$$

and from monotonicity considerations it is apparent that there is only one solution for $t > 0$; by the remarks above we know this to be at $t = 1$. If $s = -\frac{1}{2}$ we obtain the equation

$$1 - 24 \sum_{n=1}^{\infty} (-1)^n \sigma(n) e^{-2\pi nt} - 3/\pi t = 0. \tag{40}$$

The derivative with respect to t of the left side of (40) is

$$48\pi e^{-2\pi t}\,(-1 + \sum_{n=2}^{\infty}(-1)^n n\sigma(n)e^{-2\pi(n-1)t}) + 3/\pi t^2 \qquad (41)$$

and the sum from n = 2 to ∞ can be estimated as before. The same upper bound $\frac{1}{5}$ is obtained, and so (41) exceeds

$$-288\pi e^{-2\pi t}/5 + 3/\pi t^2 = (288\pi/5t^2)(15/288\pi^2 - t^2e^{-2\pi t}).$$

But the maximum of $t^2e^{-2\pi t}$ for $t \geqslant \frac{1}{2}\sqrt{3}$ occurs at $t = \frac{1}{2}\sqrt{3}$ itself; the inequality

$$15/288\pi^2 > 3\gamma/4$$

is easily checked, and it follows that the left side of (40) is monotonic. We have already found the solution $t = \frac{1}{2}\sqrt{3}$, and this completes the proof of the lemma.

3.3 Proof of Theorem III

We complete the demonstration of this theorem by proving the linear independence of the four numbers 1, ω_1, η_1 and $2\pi i$. From the transcendence of ω_1/π and ω_1/η_1 and the result of Coates in [12] it will suffice to show that the relation

$$\alpha\omega_1 + \beta\eta_1 + \gamma.2\pi i = 0 \qquad (42)$$

is impossible with $\frac{1}{4}g_2$, $\frac{1}{4}g_3$, α, $\beta \neq 0$, $\gamma \neq 0$ all algebraic integers. Accordingly we proceed to deduce a contradiction from the hypothesis that (42) holds.

For a large number k we set

$$L = [k^{4/5}] \quad , \quad h = [k^{1/10}]$$

and

$$f(z_1,z_2) = \alpha\omega_1 z_1 + \beta\zeta(\omega_1 z_1) + \gamma.2\pi i z_2 \ ;$$

then we write

$$\Phi(z_1,z_2) = \sum_{\lambda_0=0}^{L}\sum_{\lambda_1=0}^{L}\sum_{\lambda_2=0}^{L} p(\lambda_0,\lambda_1,\lambda_2)(f(z_1,z_2))^{\lambda_0}(\wp(\omega_1 z_1))^{\lambda_1}e^{2\pi i\lambda_2 z_2},$$

where the coefficients $p(\lambda_0,\lambda_1,\lambda_2)$ are chosen by the following lemma.

Lemma 3.3

There exist rational integers $p(\lambda_0,\lambda_1,\lambda_2)$, not all zero, with absolute values at most k^{10k}, such that

$$\Phi_{m_1,m_2}(s+\tfrac{1}{2},s+\tfrac{1}{2}) = 0$$

for all integers s with $1 \leqslant s \leqslant h$ and all non-negative integers m_1, m_2 with $m_1 + m_2 \leqslant k$.

Proof

Since the parameters are the same as those of the auxiliary function of Chapter II, the estimates in the proof of Lemma 2.5 are valid a fortiori; hence that proof remains virtually unchanged in the present circumstances.

We now need an analogue of Lemma 2.9, and to state it we redefine K as k^{10}.

Lemma 3.4

Let z be a complex number congruent to a point of $\mathfrak{D}$ modulo the periods of $\wp(\omega_1 z)$ whose absolute value does not exceed K. Then for all integers m with $0 \leqslant m \leqslant L$ we have

$$|\Phi_{0,m}(z,z)| < e^{-K^4}.$$

Proof

The conditions on z are enough to ensure that the extrapolation techniques of Chapter II are applicable. Consequently the proof closely follows the proofs of Lemma 2.8 and Lemma 2.9, except that the range of values of J in Lemma 2.8 is now unnecessarily large.

We now construct the Wronskian with respect to z_2 as

in section 2.5; the corresponding function $\Psi(z_1,z_2)$ takes the form

$$\Psi(z_1,z_2) = \sum_{\nu_0=0}^{N} \sum_{\nu_1=0}^{N} r(\nu_0,\nu_1)(f(z_1,z_2))^{\nu_0}(\wp(\omega_1 z_1))^{\nu_1},$$

where $N = L(L + 1)$ and the coefficients $r(\nu_0,\nu_1)$ are algebraic integers of $\mathbb{Q}(\gamma)$ with size at most $c_1^{k^2}$. In view of the remark at the end of section 2.5 the vanishing of these coefficients will prove the Theorem since the functions $f(z_1,z_2)$ and $\wp(\omega_1 z_1)$ are plainly algebraically independent.

To show that $r(\nu_0,\nu_1) = 0$ for all ν_0, ν_1 we follow the proof of Lemma 2.10 and conclude that for all z in the range of Lemma 3.4 we have

$$|\Psi(z,z)| < e^{-K^2}.$$

We use this inequality at the points

$$z = z(\ell,m) = \tfrac{1}{4} + \ell N^{-2} + m\tau \qquad (0 \leqslant \ell,m \leqslant N).$$

If we fix ℓ and write

$$s(\nu_1) = \sum_{\nu_0=0}^{N} r(\nu_0,\nu_1)(y(\ell))^{\nu_0}$$

where

$$y(\ell) = \wp(\tfrac{1}{4}\omega_1 + \omega_1 \ell N^{-2}),$$

it is clear that

$$\Psi(z(\ell,m),z(\ell,m)) = \sum_{\nu_1=0}^{N} s(\nu_1)(x(m))^{\nu_1}$$

where

$$x(m) = f(z(\ell,m),z(\ell,m)).$$

From (42) and the Legendre relation we see that

$$x(m) = x(0) - 2\pi i m\beta/\omega_1,$$

and hence from Lemma 1.3

$$|s(\nu_1)| \leqslant e^{-2K} \qquad (0 \leqslant \nu_1 \leqslant N).$$

Finally for all ν_0, ν_1

$$|r(\nu_0,\nu_1)| \leqslant e^{-K}$$

which implies $r(\nu_0,\nu_1) = 0$ as before. The argument now proceeds as in section 2.5 and this completes the proof of Theorem III.

It is interesting to note that the transcendence of $\omega_1{}^2/\pi$ in the case of complex multiplication follows from the identity

$$2\pi iC/\omega_1{}^2 = (B + 2C\tau)\eta_1/\omega_1 + \kappa$$

and the transcendence of η_1/ω_1 (cf. a remark of Lang, [21] p.652).

We conclude with a corollary referring to an arbitrary Weierstrass elliptic function with algebraic invariants; even this appears to be new.

Corollary

The ratio of the quasi-periods is transcendental if and only if neither invariant vanishes.

Proof

If $\wp(z)$ has no complex multiplication this is clear from Theorem II. Otherwise if $\wp$ has complex multiplication and $\beta = \eta_2/\eta_1$ is an algebraic number, the relation $\beta\eta_1 - \eta_2 = 0$ must be linearly dependent on the two relations already constructed, and from (36) this involves $\kappa = 0$. The corollary now follows from Lemma 3.2.

CHAPTER FOUR

In [13] Coates proved the linear independence over $\mathbb{A}$ of the numbers ω_1, ω_2 and $2\pi i$ when $\wp$ has no complex multiplication. But, as remarked in the Introduction, this proof appeals to the deep result of Serre asserting that the field of division-values of order ℓ has degree at least $c_1\ell^4$ over $\mathbb{Q}$. Moreover the constant c_1 may fail to be effectively computable in terms of the invariants g_2 and g_3; for example if

$$j = 1728g_2^3/(g_2^3 - 27g_3^2)$$

is an integer. Hence it is of some interest to avoid the use of this theory, and in this brief chapter we indicate how this may be done.

For the most part we use the methods and notation of Coates' paper [13]. Thus we assume that $\frac{1}{4}g_2$ and $\frac{1}{4}g_3$ are algebraic integers and

$$\alpha_1\omega_1 + \alpha_2\omega_2 = 2\pi i$$

for algebraic numbers α_1, α_2, and for a large integer k we set

$$L = [k^{3/4}] \ , \quad h = [k^{1/10}] \ .$$

We write

$$f(z_1,z_2) = \exp(\alpha_1\omega_1 z_1 + \alpha_2\omega_2 z_2)$$

and

$$\Phi(z_1,z_2) = \sum_{\lambda_0=0}^{L}\sum_{\lambda_1=0}^{L}\sum_{\lambda_2=0}^{L} p(\lambda_0,\lambda_1,\lambda_2)(f(z_1,z_2))^{\lambda_0}(\wp(\omega_1 z_1))^{\lambda_1}(\wp(\omega_2 z_2))^{\lambda_2}.$$

We follow [13] up to and including Lemma 8, and then we proceed as follows.

Lemma 4.1

For all complex numbers z in $\mathcal{E}$ with absolute value at most k^3 and all integers m with $0 \leqslant m \leqslant L$ we have

$$|\Phi_{o,m}(z,z)| < e^{-k^{10}}.$$

Proof

This is a little different from the proofs of Lemma 2.9 and Lemma 3.4. In the notation of Lemma 8 of [13] we write for brevity

$$Q = Q_{200} \quad , \quad S = S_{200} \quad , \quad T = T_{200} \quad , \quad Z = 10S$$

and

$$P(z_1,z_2) = \prod_{\Omega_1} (\omega_1 z_1 - \Omega_1)^{2L} \prod_{\Omega_2} (\omega_2 z_2 - \Omega_2)^{3L},$$

where Ω_i runs over all poles of $\wp(z)$ with $|\Omega_i| \leqslant |\omega_i|Z$ $(i = 1,2)$. Then if $0 \leqslant m \leqslant L$ the function

$$\phi(z_1,z_2) = P(z_1,z_2)\Phi_{o,m}(z_1,z_2)$$

is regular in the disc $|z_i| \leqslant Z$ $(i = 1,2)$. For $\Phi_{o,m}(z_1,z_2)$ may be written as a sum of terms of the type

$$h(z_1,z_2) = g(z_1,z_2)(\wp(\omega_2 z_2))^{t}(\wp'(\omega_2 z_2))^{t'}(\wp''(\omega_2 z_2))^{t''}$$

where $g(z_1,z_2)$ is an entire function of z_2 for fixed z_1, and from Lemma 1.2 the order of the pole at $z_2 = \Omega_2/\omega_2$ of $h(z_1,z_2)$ is at most

$$2t + 3t' + 4t'' \leqslant m + 2L \leqslant 3L.$$

Let $\psi(z) = \phi(z,z)$. Then from Lemma 8 of [13] with $J = 200$ we see that for all integers μ, q, r, s with $(r,q) = 1$

$$1 \leqslant q \leqslant Q, \quad 1 \leqslant s \leqslant S, \quad 1 \leqslant r < q, \quad 0 \leqslant \mu \leqslant T' = [\tfrac{1}{2}T]$$

we have

$$\psi_\mu(s + r/q) = 0.$$

For the left side of this equation is given by

$$\sum_{\nu=0}^{\mu} \binom{\mu}{\nu} \phi_{\nu,\mu-\nu}(s + r/q, s + r/q),$$

and the partial derivatives here vanish since they may be expanded in partial derivatives of Φ of order at most $T' + L < T$. For brevity we write

$$F(z) = \prod_{q=1}^{Q} \prod_{s=1}^{S} \prod_{\substack{r=1 \\ (r,q)=1}}^{q} (z - s - r/q)^{T'},$$

so that $\psi(z)/F(z)$ is regular in the disc $|z| \leqslant 5S$. Hence, denoting by θ and Θ the upper bound of $|\psi(z)|$ and the lower bound of $|F(z)|$ respectively on the circle $|z| = 5S$ we conclude from the maximum modulus principle that for any point z' in $\mathcal{E}$ with $|z'| \leqslant k^3$

$$|\psi(z')| \leqslant |F(z')|\theta/\Theta.$$

As in (14) and (15) of [13] we have

$$\theta < S^{c_2 L S^2}, \quad \Theta > 2^{c_3 Q^2 S T}|F(z')|$$

whence

$$|\psi(z')| < 2^{-c_4 Q^2 S T} < e^{-k^{10}}.$$

Finally since z' lies in $\mathcal{E}$ we have $|P(z',z')| > 1$ as usual, and the lemma follows.

We proceed to construct the Wronskian by writing

$$F(\lambda_0) = F(\lambda_0, z_1, z_2) = \sum_{\lambda_1=0}^{L} \sum_{\lambda_2=0}^{L} p(\lambda_0,\lambda_1,\lambda_2)(\wp(\omega_1 z_1))^{\lambda_1}(\wp(\omega_2 z_2))^{\lambda_2}$$

$$(0 \leqslant \lambda_0 \leqslant L)$$

and we assume the existence of an integer $M \geq 0$ such that precisely $M + 1$ of the functions $F(\lambda_0)$ are not identically zero. Denoting these by $F(r_0), \ldots, F(r_M)$, we define for $0 \leq \nu \leq M$

$$\Phi(\nu) = \Phi(\nu, z_1, z_2) = \sum_{\mu=0}^{M} (r_\mu \alpha_2 \omega_2)^\nu (f(z_1, z_2))^{r_\mu} F(r_\mu),$$

and we use Lemma 2.1 to calculate the Wronskian $W(z_1, z_2)$ of $\Phi(0), \ldots, \Phi(M)$ with respect to z_2. After some reduction we obtain the result

$$W(z_1, z_2) = (\alpha_2 \omega_2^2)^{\frac{1}{2}M(M+1)} \prod_{\lambda > \mu} (r_\lambda - r_\mu)(f(z_1, z_2))^r U(z_1, z_2)$$

where

$$r = \sum_{\mu=0}^{M} r_\mu, \quad U(z_1, z_2) = \det_{0 \leq \mu, \nu \leq M} G(\mu, \nu),$$

and

$$G(\mu, \nu) = G(\mu, \nu, z_1, z_2) = \sum_{\lambda_1=0}^{L} \sum_{\lambda_2=0}^{L} \sum_{m=0}^{\nu} \binom{\nu}{m} (r_\mu \alpha_2)^{\nu-m} p(r_\mu, \lambda_1, \lambda_2) (\wp(\omega_1 z_1))^{\lambda_1} \wp(\omega_2 z_2, \lambda_2, \nu).$$

On using Lemma 1.2 to simplify the last term in the definition of $G(\mu, \nu)$ we see that

$$U(z_1, z_2) = \sum q (\wp(\omega_1 z_1))^{\nu_1} (\wp(\omega_2 z_2))^{\nu_2} (\wp'(\omega_2 z_2))^{\nu_2'} (\wp''(\omega_2 z_2))^{\nu_2''},$$

where $q = q(\nu_1, \nu_2, \nu_2', \nu_2'')$ and the summation is over all non-negative integers $\nu_1, \nu_2, \nu_2', \nu_2''$ at most $N = kL$. Furthermore there is a positive integer a, independent of $\nu_1, \nu_2, \nu_2', \nu_2''$, such that a and aq are algebraic integers of the field $\mathbb{Q}(g_2, g_3, \alpha_2)$ with size at most $c_5^{k^2}$. Since $\wp(\omega_2 z_2)$ is an even function of z_2, it follows that the function

$$V(z_1, z_2) = U(z_1, z_2) U(z_1, -z_2)$$

may be written in the form

$$V(z_1,z_2) = \sum_{\rho_1=0}^{R} \sum_{\rho_2=0}^{R} r(\rho_1,\rho_2)(\wp(\omega_1 z_1))^{\rho_1}(\wp(\omega_2 z_2))^{\rho_2}$$

where $R = k^2$, and there exists a positive integer b such that b and $br(\rho_1,\rho_2)$ are algebraic integers of the above field with size at most $c_6^{k^2}$.

Suppose now that z is any complex number of $\mathcal{E}$ with $|z| \leqslant k^3$. We use Lemma 4.1 to estimate $|W(z,z)|$ by its first column, and we find that

$$|W(z,z)| < e^{-k^9}$$

whence

$$|V(z,z)| < e^{-k^8}.$$

Lemma 1.5 now provides an upper bound for the coefficients $r(\rho_1,\rho_2)$; the quantity μ clearly exceeds $e^{-(\log k)^4}$ by the arguments of Lemma 2.2 and hence

$$|r(\rho_1,\rho_2)| < e^{-k^7} \qquad (0 \leqslant \rho_1,\rho_2 \leqslant R).$$

Therefore $r(\rho_1,\rho_2) = 0$ for all ρ_1, ρ_2, and so $W(z_1,z_2)$ is identically zero. The argument now proceeds on lines established in section 2.5, and the proof of the result is easily completed.

CHAPTER FIVE

5.1 Introduction

In order to obtain quantitative refinements of Theorem II and Theorem III, it is natural to consider the linear form

$$\Lambda = \alpha_0 + \alpha_1\omega_1 + \alpha_2\omega_2 + \beta_1\eta_1 + \beta_2\eta_2 + \gamma.2\pi i$$

where α_0, α_1, α_2, β_1, β_2 and γ are algebraic numbers not all zero. The problem is then to find an effective lower bound for $|\Lambda|$ in terms of the maximum d of the degrees and the maximum H of the heights of the coefficients. Partial results for the case $\beta_1 = \beta_2 = \gamma = 0$ occur in a paper [4] of Baker and in our Theorem I; these take the form

$$|\Lambda| > C\exp(-(\log H)^{\kappa}) \qquad (43)$$

for some absolute constant $\kappa > 1$ and some number $C > 0$ depending only on ω_1, ω_2 and d. In view of the work of the preceding chapters the general problem may now be solved; but unless $\gamma = 0$ the actual lower bounds obtained for the case $\alpha_0 = 0$ turn out to be much weaker than (43). However, if $\alpha_0 \neq 0$, a stronger result than (43) can be derived, and the purpose of this chapter is to prove the following theorem.

Theorem IV

Let α_0, α_1, α_2, β_1, β_2 and γ be algebraic numbers with degrees at most d and heights at most H, and suppose $\alpha_0 \neq 0$. Then for any $\varepsilon > 0$ there is a constant $C > 0$ depending only on ω_1, ω_2, d and ε such that

$$|\Lambda| > C\exp(-\log H(\log\log H)^{7+\varepsilon}).$$

In fact if $\alpha_2 = \beta_1 = \beta_2 = 0$ then $7+\varepsilon$ may be replaced by $4+\varepsilon$ and in this way we obtain quite a sharp transcendence measure for the number $\pi+\omega$ where ω is any period of $\wp(z)$.

5.2 The Auxiliary Function

For the proof of Theorem IV we assume algebraic numbers $\alpha_0 \neq 0$, α_1, α_2, β_1, β_2 and γ exist with degrees at most d and heights at most $H \geqslant 3$ such that for some ε with $0 < \varepsilon < 1$

$$|\alpha_1\omega_1 + \alpha_2\omega_2 + \beta_1\eta_1 + \beta_2\eta_2 + \gamma.2\pi i - \alpha_0| < \exp(-\log H(\log\log H)^{7+\varepsilon}). \qquad (44)$$

We denote by $c, c_1, \ldots$ positive constants depending only on ω_1, ω_2, d and ε, and we shall deduce a contradiction if $H > c$ for some sufficiently large c. It is easy to see that we may take $\frac{1}{4}g_2$, $\frac{1}{4}g_3$ to be algebraic integers without loss of generality.

Let $\delta = \varepsilon/30$, and define

$$k = [\log H(\log\log H)^{2+9\delta}], \quad \ell = \log k, \quad h = [\ell^{1+\delta}]$$

$$L_0 = [\ell^{3+7\delta}], \quad L = L_1 = L_2 = [k\ell^{-1-2\delta}], \quad L_3 = [k\ell^{-\delta}].$$

We write

$$f(z_1,z_2,z_3) = \alpha_1\omega_1 z_1 + \alpha_2\omega_2 z_2 + \beta_1\zeta(\omega_1 z_1) + \beta_2\zeta(\omega_2 z_2) + \gamma.2\pi i z_3$$

and

$$\Phi(z_1,z_2,z_3) = \sum_{\lambda_0=0}^{L_0}\sum_{\lambda_1=0}^{L_1}\sum_{\lambda_2=0}^{L_2}\sum_{\lambda_3=0}^{L_3} p(\lambda_0,\lambda_1,\lambda_2,\lambda_3)(f(z_1,z_2,z_3))^{\lambda_0} (\wp(\omega_1 z_1))^{\lambda_1}(\wp(\omega_2 z_2))^{\lambda_2} e^{2\pi i\lambda_3 z_3}$$

where the coefficients $p(\lambda_0,\lambda_1,\lambda_2,\lambda_3)$ are yet to be determined.

Apart from the change in parameters, this is the same auxiliary function as that appearing in Coates' paper [12]. Hence from the calculations of [12] we see that for non-negative integers m_1, m_2, m_3

$$\Phi_{m_1,m_2,m_3}(z_1,z_2,z_3) = \omega_1^{m_1}\omega_2^{m_2}(2\pi i)^{m_3} \sum_{\lambda_0=0}^{L_0} \cdots \sum_{\lambda_3=0}^{L_3} \sum_{\mu_1=0}^{m_1} \cdots \sum_{\mu_3=0}^{m_3} p(\lambda_0, \ldots, \lambda_3) P_0 P_1 P_2 P_3,$$

where

$$P_i = \binom{m_i}{\mu_i}\wp(\omega_i z_i, \lambda_i, m_i-\mu_i), \quad P_3 = \binom{m_3}{\mu_3}\lambda_3^{m_3-\mu_3} e^{2\pi i\lambda_3 z_3}$$

for $i = 1,2$ and

$$P_0 = \sum_1 \sum_2 u_1u_2v_1v_2\tau_2(\tau_2-1)\ldots(\tau_2-\mu_3+1)\gamma^{\mu_3}\beta_1^{\lambda_0-\tau_1-\tau_1'}\beta_2^{\tau_1-\tau_2-\tau_2'}(f(z_1,z_2,z_3))^{\tau_2-\mu_3}.$$

In the formula for P_0 the numbers

$$u_1 = u_1(\tau_1,\tau_1',t_1,t_1',t_1'',\mu_1,\lambda_0),$$

$$u_2 = u_2(\tau_2,\tau_2',t_2,t_2',t_2'',\mu_2,\tau_1)$$

are rational integers with absolute values at most $\mu_1!c_1^{\mu_1+\lambda_0}$ and $\mu_2!c_2^{\mu_2+\tau_1}$ respectively, while

$$v_i = (\alpha_i - \beta_i\wp_i)^{\tau_i'}\wp_i^{t_i}\wp_i'^{t_i'}\wp_i''^{t_i''} \qquad (i = 1,2)$$

with

$$\wp_i = \wp(\omega_i z_i), \quad \wp_i' = \wp'(\omega_i z_i), \quad \wp_i'' = \wp''(\omega_i z_i).$$

Finally the summation $\sum_1$ extends over all non-negative integers τ_1, τ_1', t_1, t_1', t_1'' with

$$\tau_1 + 2\tau_1' + 2t_1 + 3t_1' + 4t_1'' = \mu_1 + \lambda_0,$$

$$\tau_1 + \tau_1' \leqslant \lambda_0$$

and $\sum_2$ over all non-negative integers τ_2, τ_2', t_2, t_2', t_2'' with

$$\tau_2 + 2\tau_2' + 2t_2 + 3t_2' + 4t_2'' = \mu_2 + \tau_1,$$

$$\tau_2 + \tau_2' \leqslant \tau_1.$$

Suppose now that s, r, q are integers with $q \neq 0$, q even and $(r,q) = 1$. We write for brevity

$$\Phi = \Phi(m_1, m_2, m_3, s, r, q) = \Phi_{m_1, m_2, m_3}(s+r/q, s+r/q, s+r/q)$$

and

$$A = A(m_1, m_2, m_3, s, r, q) = \sum_{\lambda_0=0}^{L_0} \cdots \sum_{\lambda_3=0}^{L_3} \sum_{\mu_1=0}^{m_1} \cdots \sum_{\mu_3=0}^{m_3} p(\lambda_0, \ldots, \lambda_3) Q_0 Q_1 Q_2 Q_3. \quad (45)$$

Here Q_0 is obtained from P_0 by replacing $f(z_1, z_2, z_3)$ by

$$g(s,r,q) = \alpha_0(s+r/q) - \beta_1\xi(r,0,q) - \beta_2\xi(0,r,q)$$

(see Lemma 5 of [3]) and then making the substitutions $z_1 = z_2 = z_3 = s + r/q$ in v_1 and v_2. Also Q_i is obtained from P_i $(i=1,2,3)$ by the same substitutions, so that

$$Q_i = \binom{m_i}{\mu_i}\wp(r\omega_i/q, \lambda_i, m_i-\mu_i) \;, \quad Q_3 = \binom{m_3}{\mu_3}\lambda_3^{m_3-\mu_3} e^{2\pi i r\lambda_3/q}$$

for $i = 1,2$. Note that in view of (44), Φ is very close to $\omega_1^{m_1}\omega_2^{m_2}(2\pi i)^{m_3}A$ if the coefficients $p(\lambda_0, \ldots, \lambda_3)$ are not too large, and a precise form of this remark appears in the following two lemmas.

Lemma 5.1

There exist rational integers $p(\lambda_0, \ldots, \lambda_3)$, not all zero, with absolute values at most k^{10k}, such that

$$A(m_1, m_2, m_3, s, 1, 2) = 0 \quad (46)$$

for all integers s with $1 \leqslant s \leqslant h$ and all non-negative integers m_1, m_2, m_3 with $m_1 + m_2 + m_3 \leqslant k$.

Proof

First we estimate the size of the algebraic numbers Q_i $(0 \leqslant i \leqslant 3)$ when $r = 1$, $q = 2$. The sizes of α_0, α_1, α_2, β_1, β_2, γ do not exceed c_3H since their degrees are bounded, and clearly the sizes of $\wp_i$, $\wp_i'$ and $\wp_i''$ do not exceed c_4. Since the indices τ_1, τ_1', τ_2, τ_2' are at most L_0, we obtain the upper bound $c_5^{m_1+m_2+L_0} H^{L_0}$ for the sizes of v_1 and v_2. Hence, noting that the terms in P_0 vanish if $\mu_3 > \tau_2$ and that

$$L_0 \leqslant \ell^{3+7\delta} \quad , \quad \log H \leqslant 2k\ell^{-2-9\delta} \quad , \quad H^{L_0} < k^{k/16}$$

we find that the size of Q_0 is at most

$$(m_1 + m_2 + 2L_0)^{10} m_1! m_2! L_0! c_6^{m_1+m_2+L_0} (hH)^{6L_0} < k^{2k}.$$

From Lemma 1.2 we see that

$$Q_i = \binom{m_i}{\mu_i} \sum_i{}' \; u_i' \wp_{0i}^{t_{0i}} \wp_{1i}^{t_{1i}} \wp_{2i}^{t_{2i}} \qquad (i = 1,2)$$

where $\wp_{ji} = \wp^{(j)}(\tfrac{1}{2}\omega_i)$ and the summation is over all non-negative integers t_{ji} with

$$2t_{0i} + 3t_{1i} + 4t_{2i} = m_i - \mu_i + \lambda_i,$$

and u_i' denotes a rational integer with absolute value at most $(m_i - \mu_i)! c_7^{m_i-\mu_i+\lambda_i}$. Therefore the size of Q_i is at most

$$2^{m_i} (m_i + L_i + 1)^3 m_i! c_8^{m_i+L_i} < k^{2k} \qquad (i = 1,2)$$

and since

$$Q_3 = \binom{m_3}{\mu_3} \lambda_3^{m_3-\mu_3} (-1)^{\lambda_3}$$

the size of Q_3 is at most $(2L_3)^k < k^k$.

Since $\tfrac{1}{4}g_2$, $\tfrac{1}{4}g_3$ are algebraic integers, the same is true (cf. [12], p.390) of $\wp_{ji}$ ($i = 1,2$; $j = 0,1,2$) and hence of Q_1, Q_2 and Q_3. Let a_0, a_1, a_2, b_1, b_2 and c be the positive leading coefficients in the minimal equations for α_0, α_1, α_2, β_1, β_2 and γ respectively. Then since $\xi(0,1,2) = \xi(1,0,2) = 0$, the number aQ_0 is an algebraic integer,

where

$$a = (2a_0a_1a_2b_1b_2c)^{2L_0} < (2H^6)^{2L_0} < k^k.$$

Hence aQ_0, Q_1, Q_2 and Q_3 are algebraic integers of the field $\mathbb{F}$ generated over $\mathbb{Q}$ by the numbers

$$g_2\ ,\ g_3\ ,\ \alpha_0\ ,\ \alpha_1\ ,\ \alpha_2\ ,\ \beta_1\ ,\ \beta_2\ ,\ \gamma\ ,\ \wp_{ji}$$
$$(i = 1,2;\ j = 0,1,2)$$

By Lemma 1.6 there is an integral basis $w_1, \ldots, w_f$ of $\mathbb{F}$ with the sizes of w_i at most H^{c_9}, and we may write

$$aQ_0Q_1Q_2Q_3 = n_1w_1 + \ldots + n_fw_f$$

where n_i denotes a rational integer with absolute value at most k^{9k}. It follows that (46) will be satisfied if the f equations

$$\sum_{\lambda_0=0}^{L_0} \cdots \sum_{\lambda_3=0}^{L_3} \sum_{\mu_1=0}^{m_1} \cdots \sum_{\mu_3=0}^{m_3} p(\lambda_0, \ldots, \lambda_3) n_i = 0 \qquad (1 \leqslant i \leqslant f)$$

hold for all integers s with $1 \leqslant s < h$ and all non-negative integers m_1, m_2, m_3 with $m_1 + m_2 + m_3 \leqslant k$. There are $M \leqslant fh(k + 1)^3$ equations in $N = \prod_{i=0}^{3} (L_i + 1)$ unknowns $p(\lambda_0, \ldots, \lambda_3)$, and since

$$M < c_{10}\, k^3 \ell^{1+\delta}\ ,\quad N > c_{11}\, k^3 \ell^{1+2\delta}$$

we have $N > 2M$. Thus by Lemma 1.7 the equations have a non-trivial solution in rational integers with absolute values at most

$$N \sum_{\mu_1=0}^{m_1} \sum_{\mu_2=0}^{m_2} \sum_{\mu_3=0}^{m_3} |n_i| < N(k + 1)^3 k^{9k} < k^{10k}\ ,$$

and this completes the proof of the lemma.

In the next few lemmas we shall denote the number $k\ell^{5+20\delta}$ by K.

Lemma 5.2

Let m_1, m_2, m_3 be non-negative integers with $m_1 + m_2 + m_3 \leqslant k$, and let s, r, q be integers with q even, $(r,q) = 1$

$$1 \leqslant s \leqslant \ell^4 \ , \quad 1 \leqslant r < q \leqslant \ell^2 .$$

Then

$$|\omega_1^{m_1} \omega_2^{m_2} (2\pi i)^{m_3} A - \Phi| < e^{-K}$$

where

$$A = A(m_1, m_2, m_3, s, r, q) \ , \quad \Phi = \Phi(m_1, m_2, m_3, s, r, q).$$

Furthermore, either $A = 0$ or

$$|A| > k^{-c_{12} k q^4} .$$

Proof

It is clear from Lemma 2 and Lemma 3 of [12] that A is an algebraic number of degree at most $c_{13} q^4$. To estimate its size we observe that the sizes of $g(s,r,q)$ and $\wp^{(j)}(r\omega_j/q)$ $(j = 0,1,2)$ do not exceed $c_{14} H(s + q)$ and $c_{15} q^4$ respectively. As in the previous lemma we find that the size of A is at most

$$k^{c_{16} k} (c_{17} qs)^{c_{18}(k+L_0+L_1+L_2)} \prod_{i=0}^{3} (L_j + 1) < k^{c_{19} k} .$$

Similarly A has a denominator a' given by

$$a' = q^{c_{20}(k+L_0+L_1+L_2)} a < k^{c_{21} k}$$

for some integer c_{20}. The second part of the lemma now follows from the fact that if $A \neq 0$,

$$|\mathrm{Norm}(a'A)| \geqslant 1.$$

To verify the first part we note that if

$$f = f(s+r/q, s+r/q, s+r/q) \ , \quad g = g(s,r,q)$$

then f-g differs from the left side of (46) by a factor $s+r/q$. Hence, since

$$2K < \log H\ (\log\log H)^{7+\varepsilon}$$

we obtain for $0 \leqslant \lambda_0 \leqslant L_0$

$$|f^{\lambda_0} - g^{\lambda_0}| \leqslant (|f| + |g|)^{\lambda_0 - 1}\ |f - g| \leqslant (|f| + |g|)^{\lambda_0 - 1}\ (s + r/q)e^{-2K}\ . \tag{47}$$

But we have

$$|f| < 1 + |g| < H\ell^5\ ,\quad e^K > k^k > H^{4L_0}$$

whence the number on the right of (47) is at most $e^{-3K/2}$. It is clear from previous estimates that the coefficients of g^{λ_0} in A have absolute values at most $k^{c_{22}k}$. The number of these coefficients does not exceed $k^{c_{23}}$, and the lemma follows easily.

Lemma 5.3

Suppose that $Z \geqslant 6$, and let

$$\phi(z_1,z_2,z_3) = \Phi(z_1,z_2,z_3)P(z_1,z_2)$$

where

$$P(z_1,z_2) = \prod_{i=1}^{2}\ \prod_{\Omega_i}\ (\omega_i z_i - \Omega_i)^{3L}$$

and Ω_i runs over all poles of $\wp(z)$ with $|\Omega_i| \leqslant |\omega_i|Z$. Then $\phi(z_1,z_2,z_3)$ is regular in the disc $|z_i| \leqslant Z$ $(i = 1,2,3)$, and for any z with $|z| \leqslant \frac{1}{2}Z$ and any non-negative integers $m_1,\ m_2,\ m_3$ with $m_1 + m_2 + m_3 \leqslant k$ we have

$$|\phi_{m_1,m_2,m_3}(z,z,z)| < k^{12k}\ Z^{c_{24}LZ^2}\ ,$$

$$|P_{m_1,m_2}(z,z)| < k^{2k}\ Z^{c_{25}LZ^2}\ .$$

Proof

Since $L_0 \leqslant L$ all these assertions except the last inequality may be proved by the argument of Lemma 8 of [12]. Further we have

$$|P(z_1,z_2)| < Z^{c_{26}LZ^2}$$

in the disc $|z_i| \leqslant Z$ $(i = 1,2)$ whence the remaining estimate is a consequence of the Cauchy formula

$$P_{m_1,m_2}(z,z) = \frac{m_1!m_2!}{(2\pi i)^2} \int_{C_1}\int_{C_2} \frac{P(z_1,z_2)\,dz_1\,dz_2}{(z_1 - z)^{m_1+1}(z_2 - z)^{m_2+1}}$$

where C_i is the positively described unit circle in the z_i-plane with centre at z.

Lemma 5.4

Let J be an integer satisfying

$$0 \leqslant J \leqslant 30 + [8/\delta].$$

Then

$$A(m_1,m_2,m_3,s,r,q) = 0$$

for all integers s, r, q with q even, $(r,q) = 1$,

$$1 \leqslant q \leqslant 2\ell^{J\delta/8}, \quad 1 \leqslant s \leqslant \ell^{1+\delta+J\delta/4}, \quad 1 \leqslant r < q$$

and all non-negative integers m_1, m_2, m_3 satisfying

$$m_1 + m_2 + m_3 \leqslant k/2^J.$$

Proof

The lemma is valid for $J = 0$ by Lemma 5.1. We suppose that I is an integer satisfying $0 \leqslant I < 30 + [8/\delta]$, and we assume the lemma holds for $0 \leqslant J \leqslant I$. We proceed to deduce its validity for $J = I + 1$.

We define

$$Q_J = 2\ell^{J\delta/8}, \quad S_J = \ell^{1+\delta+J\delta/4}, \quad T_J = [k/2^J]$$

and we assume that there exist integers s', r', q' with q' even, $(r',q') = 1$,

$$1 \leqslant q' \leqslant Q_{I+1} \quad , \quad 1 \leqslant s' \leqslant S_{I+1} \quad , \quad 1 \leqslant r' < q' \qquad (48)$$

and non-negative integers m_1', m_2', m_3' with

$$m_1' + m_2' + m_3' \leqslant T_{I+1}$$

such that

$$A' = A(m_1',m_2',m_3',s',r',q') \neq 0.$$

Thus we have $s' + r'/q' \neq s + r/q$, and we shall derive a contradiction if the integers m_1', m_2', m_3' are supposed minimal in the usual way.

Let $Z = 10\, S_{I+1}$ and let $\phi(z_1,z_2,z_3)$ be the corresponding function defined in Lemma 5.3. We write $\psi(z) = \phi_{m_1',m_2',m_3'}(z,z,z)$. Then for integers m, s, r, q with q even, $(r,q) = 1$

$$1 \leqslant q \leqslant Q_I \; , \quad 1 \leqslant s \leqslant S_I \; , \quad 1 \leqslant r < q \; , \quad 0 \leqslant m \leqslant T_{I+1} \qquad (49)$$

the number $\psi_m(s+r/q)$ is given by

$$\sum_{\substack{\mu_1=0 \\ \mu_1+\mu_2+\mu_3=m}}^{m} \sum_{\mu_2=0}^{m} \sum_{\mu_3=0}^{m} m!(\mu_1!\mu_2!\mu_3!)^{-1}\, \phi_{m_1'+\mu_1,\,m_2'+\mu_2,\,m_3'+\mu_3}(s+r/q,s+r/q,s+r/q).$$

The partial derivatives of ϕ occurring here may be expanded as

$$\sum_{\nu_1=0}^{m_1'+\mu_1} \sum_{\nu_2=0}^{m_2'+\mu_2} \binom{m_1'+\mu_1}{\nu_1} \binom{m_2'+\mu_2}{\nu_2} P_{\nu_1,\nu_2}\, \Phi''$$

where

$$P_{\nu_1,\nu_2} = P_{\nu_1,\nu_2}(s+r/q,s+r/q)$$

and

$$\Phi'' = \Phi(m_1'+\mu_1-\nu_1 \; , \; m_2'+\mu_2-\nu_2 \; , \; m_2'+\mu_3 \; , \; s \; , \; r \; , \; q).$$

By the induction hypothesis the algebraic number corresponding to Φ'' in the sense of Lemma 5.2 vanishes, since

$$m_1' + \mu_1 - \nu_1 + m_2' + \mu_2 - \nu_2 + m_3' + \mu_3 \leqslant T_I,$$

and hence from Lemma 5.2 it follows that $|\Phi''| < e^{-K}$.

Further we have $Z < \ell^{3+10\delta}$, and so from Lemma 5.3

$$|P_{\nu_1,\nu_2}| < k^{2k} Z^{c_{25} L Z^2} < \exp(k\ell^{5+19\delta})$$

whence

$$|\psi_m(s + r/q)| < e^{-\frac{1}{2}K}$$

for the ranges indicated in (49).

For brevity we write

$$F(z) = \prod_{\substack{q=1 \\ q \text{ even}}}^{Q_I} \prod_{s=1}^{S_I} \prod_{\substack{r=1 \\ (r,q)=1}}^{q} (z - s - r/q)^{T_{I+1}} .$$

Then for the integers s', r', q' in (48) we have

$$\frac{\psi(s'+r'/q')}{F(s'+r'/q')} = \frac{1}{2\pi i} \int_C \frac{\psi(z)\ dz}{(z-s'-r'/q')F(z)} - \frac{1}{2\pi i} \sum \frac{\psi_m(s+r/q)}{m!} I(m,s,r,q) \qquad (50)$$

where C denotes the positively described circle $|z| = 5\,S_{I+1}$;

$$I(m,s,r,q) = \int_{C_{s,r,q}} \frac{(z-s-r/q)^m\ dz}{(z-s'-r'/q')F(z)}$$

where $C_{s,r,q}$ denotes the positively described circle with centre $s+r/q$ and radius Q_{I+1}^{-2}, and the summation in (50) is over the ranges (49). We proceed to obtain from (50) an upper bound for $|\psi(s'+r'/q')|$.

Let N_I be the total number of linear factors in the product F(z), so that (see [3] p.155)

$$c_{27}\, Q_I^2\, S_I\, T_{I+1} < N_I < c_{28}\, Q_I^2\, S_I\, T_{I+1} < k\ell^{5+16\delta} .$$

From Lemma 5.3 we see that for z on C

$$|\psi(z)| < k^{12k}\,(10\,S_{I+1})^{c_{29} L S_{I+1}^2}$$

and since

$$L S_{I+1}^2 \log S_{I+1} < c_{30}\, k\ell^{1+\frac{1}{2}\delta+\frac{1}{2}I\delta} \log \ell < \ell^{-\delta/4} N_I \qquad (51)$$

this does not exceed $(3/2)^{N_I}$. Also on C we have in the usual way

$$|F(z)|^{-1} \leq 2^{-N_I} |F(s'+r'/q')|^{-1} .$$

Further, if z lies in $C_{s,r,q}$ we see that

$$|z-s'-r'/q'| \geq |s'+r'/q'-s-r/q| - |z-s-r/q| > \tfrac{1}{2}Q_{I+1}^{-2}$$

since $s'+r'/q' \neq s+r/q$; similar arguments give the estimate

$$|F(z)|^{-1} \leq (2Q_{I+1}^{2})^{N_I} < e^{K/8}$$

on this circle. Clearly also

$$|F(s'+r'/q')| \leq (2S_{I+1})^{N_I} < e^{K/8} .$$

It follows from these estimates and (50) that

$$|\psi(s'+r'/q')| < c_{31}\left(\frac{3}{2}\right)^{N_I} 2^{-N_I} + c_{32} N_I Q_{I+1}^{2} e^{-\frac{1}{4}K} < \exp(-c_{33} N_I).$$

But

$$\psi(s'+r'/q') = \sum_{\mu_1=0}^{m_1'} \sum_{\mu_2=0}^{m_2'} \binom{m_1'}{\mu_1} \binom{m_2'}{\mu_2} P'_{\mu_1,\mu_2} \Phi''' ,$$

where

$$P'_{\mu_1,\mu_2} = P_{\mu_1,\mu_2}(s'+r'/q', s'+r'/q')$$

and

$$\Phi''' = \Phi(m_1'-\mu_1, m_2'-\mu_2, m_3', s', r', q').$$

By Lemma 5.2 and the minimal choice of m_1', m_2', m_3' we see that if $\mu_1 + \mu_2 > 0$ then $|\Phi'''| < e^{-\frac{1}{2}K}$, and on isolating the term Φ' with $\mu_1 = \mu_2 = 0$ it follows that

$$|\Phi'| < |P|^{-1} \exp(-c_{34} N_I)$$

where $P = P'_{0,0}$. By a familiar argument each linear factor of P exceeds 2 in absolute value with at most $c_{35} L$ exceptions, and for these exceptions we have the lower bound c_{36}/q'. Hence $|P| > 1$ and from Lemma 5.2

$$|A'| < \exp(-c_{37} N_I).$$

On the other hand we have $A' \neq 0$ by supposition, and it follows again from Lemma 5.2 that

$$|A'| > k^{-c_{38} k Q_{I+1}^4}.$$

Since

$$k Q_{I+1}^4 \log k < c_{39} k \ell^{1+\frac{1}{2}\delta+\frac{1}{2}I\delta}$$

comparison with the right hand inequality of (51) gives a contradiction, and this proves the lemma.

5.3 Proof of Theorem IV

From Lemma 5.4 it follows that

$$A(m_1, m_2, m_3, s, 1, 4) = 0 \qquad (52)$$

for all integers m_1, m_2, m_3 and s with

$$1 \leqslant s \leqslant L_0+1, \quad 0 \leqslant m_i \leqslant L_i \qquad (i = 1,2,3).$$

Now the left side of (52) may be written in the form

$$\sum_{\lambda_0=0}^{L} \sum_{\mu_1=0}^{m_1} \sum_{\mu_2=0}^{m_2} \sum_{\mu_3=0}^{m_3} \binom{m_1}{\mu_1}\binom{m_2}{\mu_2}\binom{m_3}{\mu_3} r(\lambda_0, \mu_1, \mu_2, \mu_3, s) \, q(\lambda_0, m_1-\mu_1, m_2-\mu_2, m_3-\mu_3),$$

where

$$q(\lambda_0, \nu_1, \nu_2, \nu_3) = \sum_{\lambda_1=0}^{L_1} \sum_{\lambda_2=0}^{L_2} \sum_{\lambda_3=0}^{L_3} p(\lambda_0, \ldots, \lambda_3) \wp(\tfrac{1}{4}\omega_1, \lambda_1, \nu_1) \wp(\tfrac{1}{4}\omega_2, \lambda_2, \nu_2) \lambda_3^{\nu_3} e^{\pi i \lambda_3/2}$$

and $r(\lambda_0, \mu_1, \mu_2, \mu_3, s)$ is the expression Q_0 appearing in (45) with $r = 1$, $q = 4$. Thus (52) with $m_1 = m_2 = m_3 = 0$ gives

$$\sum_{\lambda_0=0}^{L_0} r(\lambda_0, s) q(\lambda_0, 0, 0, 0) = 0 \qquad (1 \leqslant s \leqslant L_0+1) \quad (52a)$$

where

$$r(\lambda_0, s) = r(\lambda_0, 0, 0, 0, s) = (g(s, 1, 4))^{\lambda_0}$$

and

$$g(s, 1, 4) = \alpha_0(s+\tfrac{1}{4}) - \beta_1 \xi(1, 0, 4) - \beta_2 \xi(0, 1, 4).$$

Therefore the left side of (52a) is a polynomial in s of degree at most L_0 with L_0+1 zeros, and so it must vanish identically. Since $\alpha_0 \neq 0$, it follows that

$$q(\lambda_0,0,0,0) = 0 \qquad (0 \leqslant \lambda_0 \leqslant L_0).$$

Let now ν_1, ν_2, ν_3 be any integers with $0 \leqslant \nu_i \leqslant L_i$ $(i = 1,2,3)$ and suppose we have shown that $q(\lambda_0,\nu_1',\nu_2',\nu_2')$ $= 0$ for all integers λ_0, ν_1', ν_2', ν_3' with

$$0 \leqslant \lambda_0 \leqslant L_0 \ , \quad 0 \leqslant \nu_i' \leqslant L_i \ , \quad \sum_{i=1}^{3} \nu_i' < \sum_{i=1}^{3} \nu_i .$$

Then (52) with $m_i = \nu_i$ gives

$$\sum_{\lambda_0=0}^{L_0} r(\lambda_0,s)q(\lambda_0,\nu_1,\nu_2,\nu_3) = 0 \qquad (1 \leqslant s \leqslant L_0+1),$$

and it follows by the same argument as before that $q(\lambda_0,\nu_1,\nu_2,\nu_3) = 0$ for $0 \leqslant \lambda_0 \leqslant L_0$. Thus we have proved by induction that

$$q(\lambda_0,\nu_1,\nu_2,\nu_3) = 0 \qquad (0 \leqslant \lambda_0 \leqslant L_0 \ , \ 0 \leqslant \nu_i \leqslant L_i \ , \ i = 1,2,3).$$

Now there exists λ_0 such that $p(\lambda_0,\lambda_1,\lambda_2,\lambda_3) \neq 0$ for some λ_1, λ_2, λ_3. Hence the determinant Δ of the coefficients of the $(L_1+1)(L_2+1)(L_3+1)$ equations

$$q(\lambda_0,\nu_1,\nu_2,\nu_3) = 0 \qquad (0 \leqslant \nu_i \leqslant L_i, \ i = 1,2,3)$$

must vanish. But by Lemmas 6 and 7 of [3] we have

$$\Delta = \Delta_1^{(L_2+1)(L_3+1)} \, \Delta_2^{(L_3+1)(L_1+1)} \, \Delta_3^{(L_1+1)(L_2+1)}$$

where

$$\Delta_i = 2!3! \ldots L_i! (\wp'(\tfrac{1}{4}\omega_i))^{\frac{1}{2}L_i(L_i+1)} \qquad (i = 1,2),$$

$$\Delta_3 = 2!3! \ldots L_3! (-2\pi)^{\frac{1}{2}L_3(L_3+1)} .$$

Since $\wp'(\tfrac{1}{4}\omega_i) \neq 0$, it follows that $\Delta \neq 0$. This contradiction completes the proof of Theorem IV.

CHAPTER SIX

6.1 Introduction

This chapter is devoted to the presentation of a number of lemmas on elliptic functions to be used in the proof of Theorem V in the next chapter. Although we prove nothing here of much independent interest, it is likely that the results will be useful to any future work in this field. Some of these lemmas apply to an arbitrary Weierstrass elliptic function with algebraic invariants, but in general we suppose that $\wp(z)$ has complex multiplication over the complex quadratic field $\mathbb{K}$. We also assume that $\frac{1}{4}g_2$ and $\frac{1}{4}g_3$ are algebraic integers and we write $\tau' = \omega_2/\omega_1$ in slight opposition to the notation of Chapter I. We recall that an algebraic point u of $\wp(z)$ is either a pole of $\wp(z)$ or a complex number such that $\wp(u)$ is an algebraic number.

6.2 Multiplication Formulae

The first lemma slightly sharpens the estimates of Lemma 4 and Lemma 5 of [2].

Lemma 6.1

For any positive integer ℓ there are coprime polynomials $A_\ell(x)$, $B_\ell(x)$, of degrees ℓ^2 and ℓ^2-1 respectively such that

$$\wp(\ell z) = A_\ell(\wp(z))/B_\ell(\wp(z)). \tag{53}$$

Furthermore their coefficients are algebraic integers of the field $\mathbb{Q}(g_2,g_3)$ with sizes at most $c_1^{\ell^2}$ while the leading coefficient of $A_\ell(x)$ is unity, where c_1 depends only on ω_1 and ω_2.

Proof

Define the functions $\psi_\ell = \psi_\ell(x)$ by

$$\psi_1 = 1\ ,\quad \psi_2 = 2y\ ,\quad \psi_3 = 3x^4 - 6h_2x^2 - 12h_3x - h_2^2\ ,$$

$$\psi_4 = 4y(x^6 - 5h_2x^4 - 20h_3x^3 - 5h_2^2x^2 - 4h_2h_3x - 8h_3^2 + h_2^3)$$

and generally

$$\psi_{2\ell+1} = \psi_{\ell+2}\,\psi_\ell^3 - \psi_{\ell-1}\,\psi_{\ell+1}^3 \qquad (\ell > 1),$$

$$2y\psi_{2\ell} = \psi_\ell(\psi_{\ell+2}\,\psi_{\ell-1}^2 - \psi_{\ell-2}\,\psi_{\ell+1}^2) \qquad (\ell > 2)$$

where

$$y^2 = x^3 - h_2x - h_3\ ,\quad h_2 = \tfrac{1}{4}g_2\ ,\quad h_3 = \tfrac{1}{4}g_3.$$

Then it is known that (53) is valid (see [17] p.184) if

$$A_\ell(x) = x\psi_\ell^2 - \psi_{\ell-1}\,\psi_{\ell+1}\ ,\quad B_\ell(x) = \psi_\ell^2 \tag{54}$$

and that the roots of $B_\ell(x) = 0$ are the division values

$$\alpha(m_1,m_2) = \wp((m_1\omega_1 + m_2\omega_2)/\ell) \qquad (0 \leqslant m_1,m_2 < \ell),$$

for integers m_1, m_2 not both zero. The assertions about integrality and the leading coefficient are now readily verified by induction on ℓ. Also the function ϕ_ℓ given by ψ_ℓ if ℓ is odd and $y^{-1}\psi_\ell$ if ℓ is even is a polynomial

$$\phi_\ell(x) = \ell \prod (x - \alpha(m_1, m_2))$$

where the product is over integers m_1, m_2 not both zero with

$$0 \leqslant m_1 < \ell , \quad 0 \leqslant m_2 < \tfrac{1}{2}\ell.$$

To estimate the coefficients of ϕ_ℓ we note that from Lemma 1 of [2]

$$|\alpha(m_1, m_2)| < (c_2 \ell/(m_2+1))^2$$

for the zeros of ϕ_ℓ. Hence if r does not exceed the degree of ϕ_ℓ, the product of r roots of ϕ_ℓ is at most

$$(c_2\ell)^{2\ell}(c_2\ell/2)^{2\ell} \dots (c_2\ell/s)^{2\ell} = c_2^{2\ell s}(\ell^s/s!)^{2\ell} < c_2^{2\ell s} e^{2\ell^2}$$

in absolute value, where $s = 1 + [r/\ell]$; thus the absolute value of the r-th symmetric function of the roots is at most

$$\binom{\ell^2}{r} c_2^{2\ell s} e^{2\ell^2} < c_3^{\ell^2}.$$

Therefore from (54) similar upper bounds are valid for the absolute values of the coefficients of $A_\ell(x)$ and $B_\ell(x)$.

Finally any conjugate of these coefficients must derive from a pair of conjugates g_2^σ, g_3^σ of the original invariants. Since $(g_2^\sigma)^3 - 27(g_3^\sigma)^2 \neq 0$, there exists an elliptic function $\wp^\sigma(z)$ with these conjugates as invariants, and we may repeat the preceding argument with this function, since the zeros of the conjugate polynomial $B_\ell^\sigma(x)$ are the ℓ-th order division values of $\wp^\sigma(z)$.

<u>Lemma 6.2</u>

There exists an irrational number τ in $\mathbb{K}$ such that

$$\wp(\tau z) = A(\wp(z))/B(\wp(z)) \qquad (55)$$

where $A(x)$, $B(x)$ are polynomials with coefficients in the field $\mathbb{K}(g_2, g_3)$.

Proof

Since $\tau' = \omega_2/\omega_1$ lies in $\mathbb{K}$, there exist integers $A, B, C \neq 0$ such that

$$A + B\tau' + C\tau'^2 = 0.$$

Hence if $\tau = C\tau'$ we have

$$\tau\omega_1 = C\omega_2 \ , \quad \tau\omega_2 = -A\omega_1 - B\omega_2 .$$

It follows that $\wp(\tau z)$ has periods ω_1 and ω_2, and since it is an even function it may be expressed uniquely in the form (55) where $A(x)$, $B(x)$ are coprime and $A(x)$ has leading coefficient unity. Let $\mathbb{F}_1$ be the field generated over $\mathbb{F} = \mathbb{K}(g_2, g_3)$ by the coefficients of these polynomials, and let σ be a monomorphism of $\mathbb{F}_1$ fixing $\mathbb{F}$. If $A^{\sigma}(x)$, $B^{\sigma}(x)$ are the conjugates of $A(x)$, $B(x)$ we obtain from the Laurent expansion of (55) about the origin

$$\wp(\tau z) = A^{\sigma}(\wp(z))/B^{\sigma}(\wp(z)).$$

Therefore $A^{\sigma}(x) = A(x)$, $B^{\sigma}(x) = B(x)$ and since σ may be arbitrary this implies $\mathbb{F}_1 = \mathbb{F}$, which proves the lemma. We note that since the poles and zeros of $\wp(\tau z)$ may be written down immediately, the method of [29] p.448 clearly gives an effective means of determining $A(x)$ and $B(x)$.

Lemma 6.3

Let r, s be integers, not both zero, with absolute values at most S. Then there exist coprime polynomials $G(x)$, $H(x)$ of degrees at most $c_4 S^2$ such that

$$\wp((r + s\tau)z) = G(\wp(z))/H(\wp(z)).$$

Furthermore, the coefficients of these polynomials are

algebraic integers of $\mathbb{F}$ with size at most $c_5^{S^2}$.

Proof

We assume first that neither r nor s vanishes, and for brevity write $\lambda = r + s\tau$. From Lemmas 6.1 and 6.2 we may express the functions

$$f(z) = \wp(rz) \ , \quad g(z) = \wp(s\tau z)$$

as rational functions of $\wp(z)$ with the properties of the lemma, and by differentiation we can obtain similar formulae for

$$f_1(z) = \wp'(rz)/\wp'(z) \ , \quad g_1(z) = \wp'(s\tau z)/\wp'(z).$$

The addition theorem (see [29] p.441) now shows that

$$\wp(\lambda z) = -f(z) - g(z) + \tfrac{1}{4}(\wp'(z))^2(f_1(z) - g_1(z))^2/(f(z) - g(z))^2$$
$$= C(\wp(z))/D(\wp(z)),$$

where $C(x)$, $D(x)$ have degrees at most $c_6 S^2$ and have coefficients that are algebraic integers of $\mathbb{F}$ with size at most $c_7^{S^2}$. Let $E(x)$ be the highest common factor of $C(x)$ and $D(x)$ normalized so that its leading coefficient is unity. We set

$$G(x) = C(x)/E(x) \ , \quad H(x) = D(x)/E(x)$$

and proceed to verify the assertions of the lemma. For an automorphism σ of $\mathbb{C}$ let g_2^{σ}, g_3^{σ}, τ^{σ} denote a set of conjugates of g_2, g_3, τ; we apply the lemma of [18], p.14 to the equation

$$E^{\sigma}(x)G^{\sigma}(x) = C^{\sigma}(x).$$

Since (in the terminology of that lemma) the height of $E^{\sigma}(x)$ is at least unity and the degree and height of $C^{\sigma}(x)$ are at most $c_6 S^2$ and $c_7^{S^2}$ respectively, we deduce that the height of $G^{\sigma}(x)$ is at most $c_8^{S^2}$. Hence, arguing thus with each σ, we see that the sizes of the coefficients of $G(x)$ do not exceed $c_8^{S^2}$, and similarly for $H(x)$.

Finally these coefficients are algebraic integers. For from Satz 28 of [19] it follows that if the polynomial

$$C(x) = a(x - \alpha_1) \ldots (x - \alpha_n)$$

has algebraic integer coefficients so has

$$G(x) = a(x - \alpha_1) \ldots (x - \alpha_m)$$

for $1 \leqslant m \leqslant n$.

If either r or s vanishes, the lemma is proved in the same way, except that the polynomials $C(x)$ and $D(x)$ can be obtained immediately without the use of the addition theorem.

6.3 Estimates for Algebraic Points

In this section we use the lemmas of the preceding section to deduce estimates relating to the multiplication and division of algebraic points. We call an algebraic point u non-torsion if no non-zero integer multiple of it is a period of $\wp(z)$. When $\wp(z)$ has complex multiplication this is equivalent to the condition that u/ω_1 does not lie in $\mathbb{K}$.

Lemma 6.4

Let u be a non-torsion algebraic point of $\wp(z)$ such that $\wp(u)$ is an algebraic integer. Let r, s, q be integers with r, s not both zero and

$$|r|, |s| \leqslant S, \quad 1 \leqslant q \leqslant Q.$$

Then if

$$v = (r + s\tau)u/q \;, \quad \alpha = \wp(v) \;, \quad \alpha' = \wp'(v) \;, \quad \alpha'' = \wp''(v)$$

the numbers α, α', α'' lie in a field of degree at most Q^2 over $\mathbb{F}(\wp(u), \wp'(u))$ and there is a non-zero rational integer a such that

$$a \;, \; a\alpha \;, \; a\alpha' \;, \; a\alpha''$$

are algebraic integers with sizes at most $c_9^{(S+Q)^2}$, where c_9 depends on ω_1, ω_2 and u.

<u>Proof</u>

Write for brevity $\rho = (r + s\tau)/q$; thus from Lemma 6.3

$$\wp(q\rho z) = G(\wp(z))/H(\wp(z)), \tag{56}$$

and the numbers $\gamma = G(\wp(u))$, $\delta = H(\wp(u))$ are algebraic integers of the field $\mathbb{F}_u$ generated over $\mathbb{F}$ by $\wp(u)$ and $\wp'(u)$ with sizes at most $c_{10}^{S^2}$. Furthermore $\delta \neq 0$, for otherwise $q\rho u$ would be a pole of $\wp(z)$. Therefore d = Norm δ is a non-zero rational integer of absolute value at most $c_{11}^{S^2}$, and the inequality $|d| \geq 1$ shows that $|\delta^\sigma| > c_{12}^{-S^2}$ for any conjugate δ^σ of δ. It follows that $\varepsilon = \gamma/\delta$ is an algebraic number of size at most $c_{13}^{S^2}$. But from Lemma 6.1 we have

$$\wp(q\rho z) = A_q(\wp(\rho z))/B_q(\wp(\rho z)) \tag{57}$$

whence, on putting z = u in (56) and (57) we see that

$$A_q(\alpha) - \varepsilon B_q(\alpha) = 0. \tag{58}$$

Since the leading coefficient of $A_q(x)$ is unity, it easily follows that the absolute value of α is at most $c_{14}^{(S+Q)^2}$, and we obtain a similar bound for its size by taking the conjugates of (58). Also $d\alpha$ is an algebraic integer, for if we multiply (58) by d we obtain an algebraic equation for α with algebraic integer coefficients and leading coefficient d. Finally it is clear that the field $\mathbb{F}_u(\alpha)$ has degree at most Q^2 over $\mathbb{F}_u$.

Now we have from the differential equation

$$\alpha'^2 = 4(\alpha^3 - h_2\alpha - h_3) \;, \quad \alpha'' = 6\alpha^2 - 2h_2;$$

these show that we may take $a = d^2$ and that α'' lies in $\mathbb{F}_u(\alpha)$. Also from differentiating (56) and (57) and putting $z = u$ we find that

$$\wp'(q\rho u)/\wp'(u) \;, \quad \wp'(q\rho u)/\wp'(\rho u)$$

both lie in $\mathbb{F}_u(\alpha)$. Hence so does $\alpha' = \wp'(\rho u)$, and since the estimates for a, $a\alpha$, $a\alpha'$ and $a\alpha''$ are clear this completes the proof of the lemma.

The next lemma is of a rather technical nature and its purpose is to show that the size of certain algebraic numbers is frequently much smaller than a crude estimation might imply.

Lemma 6.5

Let $u_1, \ldots, u_n$ be non-torsion algebraic points of $\wp(z)$. Then for any number $S \geq 1$ at least $\frac{1}{2}S^2$ of the numbers $r + s\tau$ ($|r|,|s| \leq S$) are such that the size of $\wp((r + s\tau)u_i)$ does not exceed c_{16} for $1 \leq i \leq n$, where c_{16} depends only on $u_1, \ldots, u_n$, ω_1 and ω_2.

Proof

Let $\mathbb{L}$ denote the field generated over $\mathbb{F}$ by the numbers $\wp(u_i)$ $(1 \leq i \leq n)$. We write $t = 1 + [4\sqrt{n\ell}]$ where $\ell = [\mathbb{L} : \mathbb{Q}]$

and for each pair of integers p, q not both zero we define the range $\mathcal{R}_{p,q}$ to consist of all pairs of integers r, s with

$$pt \leqslant r < (p + 1)t \ , \quad qt \leqslant s < (q + 1)t.$$

We first show that there exists a constant c_{16} such that the inequality

$$|\wp((r + s\tau)u_1)| \geqslant \mu \qquad (59)$$

has at most one solution in $\mathcal{R}_{p,q}$ for all p, q, provided $\mu \geqslant c_{16}$. For if (59) holds with two distinct pairs r_1, s_1 and r_2, s_2 in $\mathcal{R}_{p,q}$ it follows that poles Ω_1, Ω_2 of $\wp(z)$ exist with

$$|(r_i + s_i\tau)u_1 - \Omega_i| \leqslant c_{17}\mu^{-\frac{1}{2}}$$

(since the poles of $\wp$ are double). Hence

$$|(r_0 + s_0\tau)u_1 - \Omega_0| \leqslant 2c_{17}\mu^{-\frac{1}{2}} \qquad (60)$$

where

$$r_0 = r_1 - r_2 \ , \quad s_0 = s_1 - s_2 \ , \quad \Omega_0 = \Omega_1 - \Omega_2.$$

But since $0 \leqslant r_0, s_0 < t$ and by supposition u_1/ω_1 is not in $\mathbb{K}$ the left side of (60) has a positive lower bound c_{18}; thus if $2c_{17}\mu^{-\frac{1}{2}} < c_{18}$ we have the required contradiction.

Suppose now σ is any embedding of $\mathbb{L}$ into $\mathbb{C}$. Let $\wp^{\sigma}(z)$ be the Weierstrass elliptic function with invariants g_2^{σ}, g_3^{σ} and let u_i^{σ} be any complex number such that $\wp^{\sigma}(u_i^{\sigma}) = (\wp(u_i))^{\sigma}$. Then $\wp^{\sigma}((r + s\tau^{\sigma})u_i^{\sigma})$ is a conjugate of $\wp((r + s\tau)u_i)$. For there is a rational function $\Phi(w,x,y,z)$ with rational coefficients such that

$$\wp((r + s\tau)z) = \Phi(g_2, g_3, \tau, \wp(z))$$

from Lemma 6.3; by operating on the Laurent expansions of

both sides about the origin with σ we deduce that

$$\wp^{\sigma}((r + s\tau^{\sigma})z) = \Phi(g_2^{\sigma}, g_3^{\sigma}, \tau^{\sigma}, \wp^{\sigma}(z)),$$

and the above assertion follows on putting $z = u_i^{\sigma}$. In particular u_i^{σ} is a non-torsion algebraic point of $\wp^{\sigma}(z)$. Hence we may repeat the argument of the first paragraph to conclude that there is at most one pair r,s in $\mathcal{R}_{p,q}$ such that

$$|(\wp((r + s\tau)u_i))^{\sigma}| \geqslant c_{19}$$

for some c_{19} independent of i or σ.

Thus there are at most $n\ell$ pairs in $\mathcal{R}_{p,q}$ such that the size of $\wp((r + s\tau)u_i)$ exceeds c_{19} for some i. But there are $t^2 > 4n\ell$ pairs in $\mathcal{R}_{p,q}$; therefore at least three-quarters of these pairs fail to possess this property simultaneously for $i = 1,2, \ldots ,n$. This plainly implies the assertion of the lemma if S is large enough, and if S is small we may increase c_{16} to ensure its general validity.

The next lemma is a fairly straightforward extension of Kronecker's Theorem dealing with diophantine approximation in the fundamental parallelogram of $\wp(z)$. We denote by Λ the lattice in the complex z-plane with basis 1 and τ, and for any complex number z we write $||z||$ for the distance from z to the nearest point of Λ. Also we split z into real components z', z" with respect to Λ by writing $z = z' + \tau z''$.

Lemma 6.6

Let $v_1, \ldots ,v_n$ be complex numbers such that $v_1, \ldots ,v_n$ and 1 are linearly independent over $\mathbb{K}$, and for $X \geqslant 1$ let

$$\mu = \min \|v_1x_1 + \ldots + v_nx_n\|$$

taken over all lattice points $x_1, \ldots, x_n$ of Λ, not all zero, with absolute values at most X. Then for any complex numbers $z_1, \ldots, z_n$ the inequalities

$$\|v_ix - z_i\| < c_1X^{-1} \quad (1 \leq i \leq n) \tag{61}$$

are solvable for x in Λ with absolute value at most $c_2\mu^{-1}$.

Proof

We split the numbers v_i into components v_i', v_i'' and define the linear forms

$$L_i'(x',x'') = v_i'x'/b - v_i''x'' ,$$
$$L_i''(x',x'') = -v_i''x' - (v_i' - av_i'')x''$$

for $1 \leq i \leq n$, where a, b are rational integers such that $\tau^2 + a\tau + b = 0$ (recall from Lemma 6.2 that τ is an algebraic integer). Thus we have

$$v_ix = bL_i' - \tau L_i'' ,$$

and clearly the inequalities (61) will be satisfied if we can find integers x', x'' with absolute values at most $c_3\mu^{-1}$ such that

$$\|L_i'(x',x'') - z_i'/b\|_z < c_4X^{-1} ,$$
$$\|L_i''(x',x'') + z_i''\|_z < c_4X^{-1} \tag{62}$$

where for a real number ξ we temporarily denote the distance from ξ to the nearest rational integer by $\|\xi\|_z$. If we now define linear forms M', M'' of the variables $x_1', \ldots, x_n'$, $x_1'', \ldots, x_n''$ by

$$M' = \sum_{i=1}^{n} (v_i'x_i'/b - v_i''x_i''),$$
$$M'' = - \sum_{i=1}^{n} (v_i''x_i' + (v_i' - av_i'')x_i'')$$

we observe that

$$v_1x_1 + \dots + v_nx_n = bM' - \tau M''$$

and also that M', M'' are dual to L_i', L_i''. Then from the definition of μ we have for any real c

$$\max\{c\mu^{-1}\max(\|M'\|_z, \|M''\|_z), cX^{-1}\max_i \max(|X_i'|, |X_i''|)\} \geqslant cc_5$$

where x_i', x_i'' are now arbitrary rational integers not all zero. But if c is so large that

$$cc_5 > (\ell!)^2 2^{-\ell+1}, \quad \ell = 2n + 2$$

we see from [9], p.99, Theorem XVIIB that the inequalities (62) are solvable in rational integers x', x'' with the required upper bounds. This proves the lemma.

The next result is that particular consequence of Lemma 6.6 which will be used in the proof of Theorem V. For a positive integer m we denote by Λ/m the lattice with basis m^{-1} and $m^{-1}\tau$. We recall the definition of the unit ball $\mathbb{B}$ in $\mathbb{C}^n$ as the set of points $\underline{z} = (z_1, \dots, z_n)$ with

$$\underline{z} = (|z_1|^2 + \dots + |z_n|^2)^{1/2} \leqslant 1.$$

<u>Lemma 6.7</u>

Let $u_1, \dots, u_n$ be algebraic points of $\wp(z)$ linearly independent over $\mathbb{K}$. For an integer $k \geqslant 1$ we write

$$\mu(k) = \min |u_1x_1 + \dots + u_nx_n|$$

taken over all lattice points $x_1, \dots, x_n$ of Λ, not all zero, with absolute values at most k^5. Then for any point $(\xi_1, \dots, \xi_n)$ of $\mathbb{B}$ there are at least k distinct points x of Λ/k^5 such that

$$|\wp(u_i x) - \xi_i| < k^{-3} \quad , \quad |x| < c_7(\mu(k))^{-1} \tag{63}$$

where c_7 depends only on $u_1, \ldots, u_n$, ω_1 and ω_2.

<u>Proof</u>

We shall assume k is large and we shall construct solutions of (63). We write $v_i = u_i/u_n$ for $1 \leqslant i \leqslant n-1$. Then for lattice points $x_1, \ldots, x_{n-1}$ of Λ, not all zero, with absolute values at most k^4 we have

$$|u_n| . \|x_1 v_1 + \ldots + x_{n-1} v_{n-1}\| \geqslant \mu = \mu(k),$$

since if x_n is the point of Λ nearest $x_1 v_1 + \ldots + x_{n-1} v_{n-1}$ we have $|x_n| < k^5$. Thus if ζ_i is any complex number (in a fixed fundamental parallelogram) such that $\wp(\zeta_i) = \xi_i$ and we put

$$z_i = (\zeta_i u_n - \zeta_n u_i)/u_n \omega_1 \qquad (1 \leqslant i \leqslant n-1)$$

we may use Lemma 6.6 to solve the inequalities

$$\|v_i x - z_i\| < c_8 k^{-4} \qquad (1 \leqslant i \leqslant n-1)$$

for x in Λ with $|x| < c_9 \mu^{-1}$. Hence, putting

$$z = (\zeta_n + x\omega_1)/u_n$$

we see that

$$|z| < c_{10}\mu^{-1} \quad , \quad \|u_i z/\omega_1 - \zeta_i/\omega_1\| < c_{11} k^{-4} \qquad (1 \leqslant i \leqslant n). \tag{64}$$

Finally we let x_0 be the point of Λ/k^5 nearest to z and we write for $0 \leqslant r < k$

$$x_r = x_0 + r/k^5$$

so that

$$|x_r| < c_{12}\mu^{-1} \quad , \quad |x_r - z| < c_{13} k^{-4} .$$

We proceed to verify (63) for $x = x_r$. There is a period Ω_i, from (64), such that

$$|u_i x - \zeta_i - \Omega_i| < c_{14} k^{-4} . \tag{65}$$

Since $|\wp(\zeta_i)| \leqslant 1$ there is a circle C of radius c_{15} with centre ζ_i all of whose points are at least c_{15} from each pole of $\wp(z)$. From (65) we deduce that the point $\eta_i = u_i x - \Omega_i$ is inside C and at least $\frac{1}{2}c_{15}$ from each point of C. Hence in the formula

$$\wp(\eta_i) - \wp(\zeta_i) = \frac{(\eta_i - \zeta_i)}{2\pi i} \int_C \frac{\wp(z)\ dz}{(z - \eta_i)(z - \zeta_i)}$$

the absolute value of the integrand is at most c_{16}, and the inequalities (63) are now apparent. If k is not large we can adjust c_7 to ensure the universal validity of the lemma.

CHAPTER SEVEN

7.1 Introduction

In this chapter we prove the linear forms theorem promised in the general introduction. We suppose that $\wp(z)$ is a Weierstrass elliptic function with complex multiplication over $\mathbb{K}$, and we let $u_1, \ldots ,u_n$ be algebraic points of $\wp(z)$ that are linearly independent over $\mathbb{K}$.

Theorem V

For any $\varepsilon > 0$ and any positive integer d there is a constant $C > 0$ effectively computable in terms of g_2, g_3, $u_1, \ldots , u_n$, d and ε such that

$$|\alpha_1 u_1 + \ldots + \alpha_n u_n| > Ce^{-H^{\varepsilon}} \qquad (66)$$

for all algebraic numbers $\alpha_1, \ldots ,\alpha_n$, not all zero, with degrees at most d and heights at most H.

Before we can prove this theorem we need to establish four lemmas on apparently miscellaneous topics.

7.2 Four Lemmas

The first of these replaces, largely for reasons of elegance and brevity, the kind of result expressed by Lemmas 1.9, 2.6 and 5.3. We denote by $\sigma(z)$ the entire function with simple zeros at the poles of $\wp(z)$. Then we have (see [29], p.447)

$$\sigma(z) = z \prod_{\Omega} (1 - z/\Omega) \exp(z/\Omega + \tfrac{1}{2}z^2/\Omega^2)$$

where the product runs over all non-zero poles of $\wp(z)$.

Lemma 7.1

The functions $\sigma(z)$ and $(\sigma(z))^2\wp(z)$ are entire and for $Z \geqslant 1$ their maximum moduli on $|z| \leqslant Z$ do not exceed $c_1^{Z^2}$. Furthermore if δ is the distance from z to the nearest pole of $\wp(z)$, we have for $|z| \leqslant Z$

$$|\sigma(z)| \geqslant \delta c_2^{-Z^2},$$

where c_1, c_2 depend only on ω_1 and ω_2.

Proof

Since the poles of $\wp(z)$ are double, it is clear that $f(z) = (\sigma(z))^2\wp(z)$ is an entire function. Also, from [29] p.448 we have

$$\sigma(z + \omega_i) = e^{\alpha_i z + \beta_i}\sigma(z) \qquad (i = 1,2) \tag{67}$$

for certain complex numbers α_i, β_i depending only on ω_1 and ω_2. It follows that for integers m_1, m_2

$$\sigma(z_0 + m_1\omega_1 + m_2\omega_2) = \sigma(z_0) \exp\{(m_1\alpha_1 + m_2\alpha_2)z_0 + Q(m_1,m_2)\} \tag{68}$$

where $Q(m_1,m_2)$ is a quadratic polynomial in m_1 and m_2. For any complex number z with $|z| \leqslant Z$ we define z_0 and integers m_1, m_2 by

$$z = z_0 + m_1\omega_1 + m_2\omega_2$$

where z_0 lies in the fundamental parallelogram consisting of points $\theta_1\omega_1 + \theta_2\omega_2$ with $0 \leqslant \theta_1,\theta_2 < 1$. Since ω_2/ω_1 is not real we see that

$$|m_1|, |m_2| < c_3(|z| + 1) < c_4 Z$$

and hence

$$|Q(m_1, m_2)| < c_5 Z^2,$$

whence the desired upper bound for $|\sigma(z)|$ now follows from (68) and the boundedness of $|\sigma(z_0)|$. Also by squaring (68) and multiplying both sides by $\wp(z_0) = \wp(z)$ we obtain a functional equation for f(z), and the estimate for $|f(z)|$ follows from this in a similar way.

Finally if δ is the distance from z to the nearest pole we have $|\sigma(z_0)| \geqslant c_6\delta$ since the zeros of $\sigma(z)$ are simple; from (68) again we conclude that

$$|\sigma(z)| \geqslant \delta c_2^{-Z^2}.$$

It is well-known, and easy to prove, that if ρ is a complex number such that $\wp(z)$ and $\wp(\rho z)$ are algebraically dependent, then ρ must lie in $\mathbb{K}$ (cf. Lemma 6.3, a kind of converse). The next lemma is a generalization of this remark, in which $z_1, \ldots, z_n$ are independent complex variables.

<u>Lemma 7.2</u>

There is a constant c_7 depending only on ω_1 and ω_2 with the following property. Let $\alpha_1, \ldots, \alpha_n$ be non-zero complex numbers and suppose ϕ is a rational function of total degree D in the functions

$$\wp(z_1), \ldots, \wp(z_{n-1}), \wp(\alpha_1 z_1 + \ldots + \alpha_n z_n)$$

and their first and second partial derivatives. Further suppose that ϕ is independent of $z_1, \ldots, z_{n-1}$. Then if α_1 is not in $\mathbb{K}$, ϕ must be a constant; and the same conclusion holds if α_1 is in $\mathbb{K}$ and its height exceeds c_7D.

Proof

Without loss of generality we may assume that $|\alpha_1| \leqslant 1$, for the substitutions

$$z_1' = \alpha_1 z_1 + \ldots + \alpha_n z_n\ , \quad z_i' = z_i \qquad (2 \leqslant i \leqslant n)\ ,$$
$$\alpha_1' = \alpha_1^{-1}\ , \quad \alpha_i' = -\alpha_i/\alpha_1 \qquad (2 \leqslant i \leqslant n)$$

give

$$z_1 = \alpha_1' z_1' + \ldots + \alpha_n' z_n'$$

and preserve the shape of the problem while replacing α_1 by α_1^{-1} (recall that the heights of α_1, α_1^{-1} are equal).

Let $\mathbb{E}$ denote the field of meromorphic functions of $z_1, \ldots, z_n$ that are doubly periodic in z_1 with the same periods as $\wp(z_1)$. We start by supposing that α_1 lies in $\mathbb{K}$, and has height $H > cD$ for some c which is yet to be determined. We proceed to prove that if $z_0 = \alpha_1 z_1 + \ldots + \alpha_n z_n$ the degree of $f = \wp(z_0)$ over $\mathbb{E}$ is at least c_8H.

For any positive integer M the M^2 transformations

$$z_1 \longrightarrow z_1^{\sigma} = z_1 + m_1\omega_1 + m_2\omega_2 \qquad (1 \leqslant m_1, m_2 \leqslant M)$$

give rise to automorphisms of the field of meromorphic functions in $z_1, \ldots, z_n$ that fix $\mathbb{E}$, and we assert that M can be chosen so that f has a large number of distinct conjugates f^{σ}. For the equation $f^{\sigma_1} = f^{\sigma_2}$ with $\sigma_1 \neq \sigma_2$ implies that $\alpha_1(z_1^{\sigma_1} - z_1^{\sigma_2})$ is a period of $\wp(z)$, and hence integers m_1, m_2 exist, not both zero, with absolute values at most M, such that

$$\alpha_1(m_1\omega_1 + m_2\omega_2) = n_1\omega_1 + n_2\omega_2.$$

Since $|\alpha_1| \leqslant 1$ we see that the absolute values of the integers n_1, n_2 are at most c_9M; this shows that H does not exceed $c_{10}M^2$. But on choosing M such that

$$c_{10}M^2 < H < 2c_{10}M^2$$

we derive a contradiction. Therefore f has M^2 distinct conjugates over $\mathbb{E}$, which implies that its degree over $\mathbb{E}$ is not less than $M^2 > c_8H$.

By using the differential equations for $\wp(z)$ we can write

$$\phi = P(f) + \wp'(z_0)Q(f) \tag{69}$$

where $P(x)$, $Q(x)$ are rational functions of x of degrees at most $c_{11}D$ whose coefficients are rational functions of $\wp(z_1), \ldots, \wp(z_{n-1})$ and their first and second derivatives. From the original hypothesis ϕ certainly lies in $\mathbb{E}$; hence the rational function

$$R(x) = (4x^3 - g_2x - g_3)(Q(x))^2 - (P(x) - \phi)^2$$

has coefficients in $\mathbb{E}$, degree at most $c_{12}D$, and vanishes for $x = f$. If we choose $c > c_{12}/c_8$ we have $c_{12}D < c_8H$ and so the function $R(x)$ must vanish identically in x. Since $4x^3 - g_2x - g_3$ is not a square this implies $Q(x) = 0$ and $P(x)$ is independent of x. Thus from (69) ϕ is actually a rational function of $\wp(z_1), \ldots, \wp(z_{n-1})$ and their first and second derivatives, and the assertion of the lemma is clear.

Finally if α_1 is not in $\mathbb{K}$ the conjugates of f are all distinct for any M, since $\alpha_1(z_1^{\sigma_1} - z_1^{\sigma_2})$ is never a period of $\wp(z)$. Hence f is not algebraic over $\mathbb{E}$, and the function R(x) must vanish identically. The conclusion of the lemma now follows as before.

The next lemma concerns generalized Wronskians. Suppose $f_0(z_1, \ldots, z_n), \ldots, f_N(z_1, \ldots, z_n)$ are N+1 meromorphic functions and let

$$\partial = (\partial/\partial z_1)^{m_1} \ldots (\partial/\partial z_n)^{m_n}$$

be a differential operator of order

$$|\partial| = m_1 + \ldots + m_n.$$

For such operators $\partial_0, \ldots, \partial_N$ with $|\partial_i| \leqslant i$ $(0 \leqslant i \leqslant N)$ we define a matrix whose (r,s) entry is $\partial_r f_s$ $(0 \leqslant r,s \leqslant N)$, and we denote its determinant by $W(\partial_0, \ldots, \partial_N; f_0, \ldots, f_N)$. It is well-known that if W vanishes for all possible sets of differential operators then $f_0, \ldots, f_N$ are linearly dependent over $\mathbb{C}$. A refinement of this result runs as follows.

<u>Lemma 7.3</u>

Suppose all the Wronskian determinants of the functions $f_0, \ldots, f_N$ which are formed with differential operators not involving $\partial/\partial z_n$ vanish identically. Then there exist meromorphic functions $\psi_0, \ldots, \psi_N$ of z_n, not all zero, such that

$$\psi_0 f_0 + \ldots + \psi_N f_N = 0,$$

and furthermore each ψ_i may be expressed as a rational function of total degree N in the elements of the Wronskian matrices.

Proof

If N = 0 the lemma is trivial, and we proceed by induction on N. Thus we may clearly assume that there is some Wronskian determinant of the functions $f_0, \ldots, f_{N-1}$ which does not vanish identically, say

$$W_N = W(\partial_0, \ldots, \partial_{N-1}; f_0, \ldots, f_{N-1})$$

where $\partial_0, \ldots, \partial_{N-1}$ are independent of $\partial/\partial z_n$ and $|\partial_i| \leqslant i$. We write

$$W_r = W(\partial_0, \ldots, \partial_{N-1}; f_0, \ldots, f_{r-1}, f_{r+1}, \ldots, f_N) \quad (0 \leqslant r \leqslant N),$$

and we proceed to verify that the assertions of the lemma hold for $\psi_r = W_r/W_N$. First we have

$$W_N \sum_{r=0}^{N} \psi_r f_r = W_0 f_0 + \ldots + W_N f_N$$

and the right side of this vanishes since it is the development of the determinant

$$W(\partial_0, \partial_0, \partial_1, \ldots, \partial_{N-1}; f_0, \ldots, f_N)$$

by its first row.

Now let δ be an arbitrary operator of order 1 independent of $\partial/\partial z_n$, i.e. one of $\partial/\partial z_1, \ldots, \partial/\partial z_{n-1}$. We have

$$\delta W_r = \sum_{s=0}^{N-1} W_{rs} \qquad (0 \leqslant r \leqslant N) \tag{70}$$

where

$$W_{rs} = W(\partial_0, \ldots, \partial_{s-1}, \delta\partial_s, \partial_{s+1}, \ldots, \partial_{N-1}; \\ f_0, \ldots, f_{r-1}, f_{r+1}, \ldots, f_N)$$

is the determinant obtained by differentiating the (s+1)-th row of W_r. Consider the matrix $\underline{\underline{W}}$ associated with the determinant

$$W(\partial_0, \ldots, \partial_{s-1}, \partial_s, \delta\partial_s, \partial_{s+1}, \ldots, \partial_{N-1}; f_0, \ldots, f_N).$$

It is clear that $(-1)^{r+s} W_{rs}$, $(-1)^{s+t+1} W_t$ are the signed minors of $\partial_s f_r$ and $\delta\partial_s f_t$ respectively in this matrix. Hence apart from sign the expression

$$M_{rs} = W_N W_{rs} - W_r W_{Ns}$$

is the determinant of a two-rowed submatrix of the adjoint of $\underline{\underline{W}}$. But $\underline{\underline{W}}$ is a Wronskian matrix and therefore by hypothesis it is singular. This implies that the rank of its adjoint is at most 1 (see [7], p.287, ex.16) and so $M_{rs} = 0$. Thus from (70)

$$W_N^2 \delta\psi_r = W_N \delta W_r - W_r \delta W_N = \sum_{s=0}^{N-1} M_{rs} = 0$$

whence $\delta\psi_r = 0$ and consequently $\psi_0, \ldots, \psi_N$ are independent of $z_1, \ldots, z_{n-1}$. The lemma is now obvious from the original definition of these functions.

The final lemma of this section is a slight indulgence, for a weaker version follows from generalizing the result of Lemma 2.3, and this would suffice for our purposes. However, in view of Appendix III, it is of some interest to see how near elementary methods can approach best possible results.

Lemma 7.4

Let $\phi(z_1, \ldots, z_n)$ be a polynomial in n complex variables of degree at most L in each of them. Then if ϕ has a zero within $(2n^2L)^{-1}$ of each point of $\mathcal{B}$ it is identically zero.

Proof

We shall in fact prove this lemma under the weaker hypothesis that ϕ has a zero within $(2n^2L)^{-1}$ of each point of the region $\mathcal{U}$ defined by $|z_1| = \ldots = |z_n| = 1/\sqrt{n}$. We need the appropriate analogue of (17); this is due to Bernstein ([6], p.45) and states that

$$\sup_{|z|\leq 1} |f'(z)| \leq L \sup_{|z|\leq 1} |f(z)| \tag{71}$$

for polynomials $f(z)$ of degree L in the single complex variable z. We also require the inequality for $t \geq 1$

$$\sup_{\underline{z} \text{ in } t\mathcal{U}} |\phi(\underline{z})| \leq t^{nL} \sup_{\underline{z} \text{ in } \mathcal{U}} |\phi(\underline{z})|$$

for the polynomial ϕ, where $t\mathcal{U}$ is the region $|z_1| = \ldots = |z_n| = t/\sqrt{n}$. This comes from applying the maximum modulus principle to the polynomial

$$\psi(z_1, \ldots, z_n) = (z_1 \ldots z_n)^L \phi(z_1^{-1}, \ldots, z_n^{-1}).$$

To prove the lemma, suppose ϕ satisfies the weaker hypothesis and write $\delta = (2n^2L)^{-1}$ and

$$M = \sup_{\underline{z} \text{ in } \mathcal{U}} |\phi(\underline{z})|. \tag{72}$$

Let $\underline{\zeta}$ be any point of $\mathcal{U}$, and let $\underline{\sigma}$ be the zero of ϕ nearest to $\underline{\zeta}$, so that $|\underline{\zeta} - \underline{\sigma}| \leq \delta$. We use the generalization of (19)

$$\phi(\underline{\zeta}) - \phi(\underline{\sigma}) = \sum_{i=1}^{n} \int_{L_i} \phi_i(\sigma_1, \ldots, \sigma_{i-1}, z_i, \zeta_{i+1}, \ldots, \zeta_n)\, dz_i \,, \tag{73}$$

where L_i is the straight line joining σ_i to ζ_i, and $\phi_i = \partial\phi/\partial z_i$ $(1 \leq i \leq n)$. From (71) with an obvious scaling factor we see that $|\phi_i| \leq \sqrt{n}.LM$ on $\mathcal{U}$; hence on $t\mathcal{U}$ with $t = 1 + \delta\sqrt{n} < e^{n\delta}$ we have

$$|\phi_i| \leq e^{n^2\delta L}\, nLM \leq 7nLM/4. \tag{73a}$$

Since

$$|\sigma_i| \leq \delta + n^{-1/2} = tn^{-1/2}$$

the upper bound (73a) holds for the integrands in (73), and we deduce that

$$|\phi(\underline{\zeta})| \leq 7n^2\delta LM/4 \leq 7M/8.$$

Since $\underline{\zeta}$ is arbitrary this contradicts (72) unless $M = 0$, in which case ϕ is identically zero and the lemma is proved.

7.3 A Simplification

Suppose $u_1, \dots, u_n$ are algebraic points of $\wp(z)$ that are linearly independent over $\mathbb{K}$. We begin by observing that no generality is lost in proving Theorem V under the supposition that $u_1, \dots, u_n$ are all non-torsion points. For if this case of the theorem has been established and $u_1', \dots, u_n'$ are independent algebraic points with u_1' a torsion point (there can be at most one) we have

$$\alpha_1'u_1' + \dots + \alpha_n'u_n' = \alpha_1u_1 + \dots + \alpha_nu_n \tag{74}$$

where

$$u_1 = u_1' + u_2', \quad u_i = u_i' \quad (i \neq 1),$$

$$\alpha_2 = \alpha_2' - \alpha_1', \quad \alpha_i = \alpha_i' \quad (i \neq 2).$$

If d', H' denote the maximum of the degrees and heights respectively of $\alpha_1', \dots, \alpha_n'$ then the degrees and heights of $\alpha_1, \dots, \alpha_n$ do not exceed

$$d \leq d'^2, \quad H \leq c_1H'^{c_2}$$

respectively, where c_1, c_2 depend only on d' (cf. [1], p.206). Hence, noting that $u_1, \dots, u_n$ are all non-torsion points, we may apply the theorem to these with $\varepsilon = \varepsilon'/2c_2$, and we conclude that the absolute value of the right side of (74) exceeds $c_3e^{-H^{\varepsilon}}$. This is clearly at least $c_4e^{-H'^{\varepsilon'}}$ where c_3, c_4 depend only on $u_1', \dots, u_n'$, ω_1, ω_2, d' and ε'.

It now becomes convenient to formulate the following

modified version of the theorem.

Proposition

For any $\varepsilon > 0$ and any positive integer d there is a constant H_0 effectively computable in terms of g_2, g_3, $u_1, \ldots, u_n$, d and ε with the following property. For all non-zero algebraic numbers $\alpha_1, \ldots, \alpha_{n-1}$ of degrees at most d and heights at most $H \geqslant H_0$ such that either α_1 is not in $\mathbb{K}$ or its height is exactly H we have

$$|\alpha_1 u_1 + \ldots + \alpha_{n-1} u_{n-1} - u_n| > e^{-H^{\varepsilon}} .$$

We proceed to show that this implies Theorem V for the points $u_1, \ldots, u_n$. Let $\alpha_1', \ldots, \alpha_n'$ be algebraic numbers, not all zero, with degrees and heights at most d' and H' respectively. To estimate the linear form

$$\Lambda' = \alpha_1' u_1 + \ldots + \alpha_n' u_n$$

as in (66) with given $\varepsilon' > 0$ we may evidently assume that $\alpha_1', \ldots, \alpha_n'$ are all non-zero; thus we can write

$$\Lambda = -\Lambda'/\alpha_n' = \alpha_1 u_1 + \ldots + \alpha_{n-1} u_{n-1} - u_n$$

where $\alpha_i = -\alpha_i'/\alpha_n'$ $(1 \leqslant i \leqslant n-1)$. The degrees and heights of $\alpha_1, \ldots, \alpha_{n-1}$ do not exceed

$$d \leqslant d'^2 \ , \quad H \leqslant c_5 H'^{c_6}$$

resepctively, and there are now two possibilities.

(a) Not all of $\alpha_1, \ldots, \alpha_{n-1}$ are in $\mathbb{K}$. Then we may suppose α_1 is not in $\mathbb{K}$, and it follows from the Proposition with $\varepsilon = \varepsilon'/4c_6$ that

$$|\Lambda| > c_7 e^{-H_1^{\varepsilon}}$$

where $H_1 = \max(H, H_0)$. Thus $|\Lambda| > c_8 e^{-H'^{\varepsilon'/2}}$, and since

$|\alpha_n'| \geq (d'H')^{-1}$ it is clear that

$$|\Lambda'| > c_9 e^{-H'^{\varepsilon'}} .$$

(b) $\alpha_1, \ldots, \alpha_{n-1}$ are all in $\mathbb{K}$. Then we may suppose that α_1 is the coefficient of greatest height H. If $H \geq H_0$ the Proposition gives a lower bound for $|\Lambda|$ as before while if $H < H_0$ we have $\Lambda \neq 0$ by the linear independence of $u_1, \ldots, u_n$ over $\mathbb{K}$; so in this case $|\Lambda| > c_{10}$ and the theorem follows on combining these lower bounds.

Finally we note that we can take $\frac{1}{4}g_2$, $\frac{1}{4}g_3$, $\wp(u_i)$ $(1 \leq i \leq n)$ to be algebraic integers. For if we replace ω_1, ω_2 by ω_1/m, ω_2/m for some positive integer m the numbers g_2, g_3, u_i, $\wp(u_i)$ become $m^4 g_2$, $m^6 g_3$, u_i/m, $m^2\wp(u_i)$ resepctively and so the validity of (66) will not be affected provided the constant C is appropriately modified.

7.4 The Auxiliary Function

We shall establish Theorem V by induction on n. In his paper [16], Feldman announces a result which is essentially a stronger version of the theorem for n = 2. Therefore we assume the truth of the theorem for $n-1 \geq 2$ points and we shall eventually deduce a contradiction from its falsity for the n algebraic points $u_0, \ldots, u_{n-1}$. While we do this, however, we shall be developing techniques which will provide a self-contained proof for n = 2 more in the spirit of these Notes. The proof will be sketched without details in section 7.6.

Accordingly for the induction step we may assume from

the preceding section that there exist non-zero algebraic numbers $\alpha_1, \ldots, \alpha_{n-1}$ with degrees at most d and heights at most H, such that either α_1 is not in $\mathbb{K}$ or the height of α_1 is exactly H, and such that for some $\varepsilon > 0$

$$|\alpha_1 u_1 + \ldots + \alpha_{n-1} u_{n-1} - u_0| < e^{-H^\varepsilon} . \tag{75}$$

The non-torsion algebraic points $u_0, \ldots, u_{n-1}$ are linearly independent over $\mathbb{K}$ and $\frac{1}{4}g_2$, $\frac{1}{4}g_3$ as well as $\wp(u_i)$ $(0 \leq i \leq n-1)$ are algebraic integers. We denote by $c, c_1, \ldots$ constants depending only on $u_0, \ldots, u_{n-1}$, ω_1, ω_2, d and ε, and we assume that $H > c$ where c is chosen sufficiently large for the validity of the subsequent arguments.

We define the integer k by

$$k = [(\log H)^8]$$

and we write

$$L = [k^{1-1/8n}] \quad , \quad h = k^{1/8} .$$

Then for coefficients $p(\lambda_0, \ldots, \lambda_{n-1})$ that are yet to be determined we write

$$\Phi(z_1, \ldots, z_{n-1}) = \sum_{\lambda_0=0}^{L} \ldots \sum_{\lambda_{n-1}=0}^{L} p(\lambda_0, \ldots, \lambda_{n-1}) (\wp(u_1 z_1))^{\lambda_1} \ldots (\wp(u_{n-1} z_{n-1}))^{\lambda_{n-1}} (\wp(g(z_1, \ldots, z_{n-1})))^{\lambda_0}$$

where

$$g = g(z_1, \ldots, z_{n-1}) = \alpha_1 u_1 z_1 + \ldots + \alpha_{n-1} u_{n-1} z_{n-1} .$$

For non-negative integers $m_1, \ldots, m_{n-1}$ we have

$$\Phi_{m_1,\ldots,m_{n-1}}(z_1, \ldots, z_{n-1}) = u_1^{m_1} \ldots u_{n-1}^{m_{n-1}} \sum_{\lambda_0=0}^{L} \ldots \sum_{\lambda_{n-1}=0}^{L} \sum_{\mu_1=0}^{m_1} \ldots \sum_{\mu_{n-1}=0}^{m_{n-1}} p(\lambda_0, \ldots, \lambda_{n-1}) Q$$

where

$$Q = \binom{m_1}{\mu_1} \ldots \binom{m_{n-1}}{\mu_{n-1}} \alpha_1^{\mu_1} \ldots \alpha_{n-1}^{\mu_{n-1}} \wp(u_1 z_1, \lambda_1, m_1 - \mu_1) \ldots \wp(u_{n-1} z_{n-1}, \lambda_{n-1}, m_{n-1} - \mu_{n-1}) \wp(g, \lambda_0, \mu_1 + \ldots + \mu_{n-1}) .$$

For integers r, s, q with $q > 0$ and r, s not both zero we write

$$\Phi(m_1, \ldots, m_{n-1}, r, s, q) = \Phi_{m_1,\ldots,m_{n-1}}((r+s\tau)/q, \ldots, (r+s\tau)/q)$$

and

$$A(m_1, \ldots, m_{n-1}, s, r, q) = \sum_{\lambda_0=0}^{L} \cdots \sum_{\lambda_{n-1}=0}^{L} \sum_{\mu_1=0}^{m_1} \cdots \sum_{\mu_{n-1}=0}^{m_{n-1}} p(\lambda_0, \ldots, \lambda_{n-1})R,$$

where

$$R = \binom{m_1}{\mu_1} \cdots \binom{m_{n-1}}{\mu_{n-1}} \alpha_1^{\mu_1} \cdots \alpha_{n-1}^{\mu_{n-1}} \wp((r+s\tau)u_1/q, \lambda_1, m_1-\mu_1) \cdots$$
$$\wp((r+s\tau)u_{n-1}/q, \lambda_{n-1}, m_{n-1}-\mu_{n-1}) \wp((r+s\tau)u_0/q, \lambda_0, \mu_1+\ldots+\mu_{n-1}).$$

Thus from (75) we see that if the coefficients $p(\lambda_0, \ldots, \lambda_{n-1})$ are relatively small integers, A is an algebraic number closely related to Φ on the diagonal $z_1 = \ldots = z_{n-1} = (r+s\tau)/q$. We proceed to make this precise in the following two lemmas.

Lemma 7.5

There exist rational integers $p(\lambda_0, \ldots, \lambda_{n-1})$, not all zero, with absolute values at most c_1^{hk}, such that

$$A(m_1, \ldots, m_{n-1}, r, s, 1) = 0 \qquad (76)$$

for all integers r,s , not both zero, with absolute values at most h and all non-negative integers $m_1, \ldots, m_{n-1}$ with

$$m_1 + \ldots + m_{n-1} \leqslant k.$$

Proof

We examine the algebraic number R when $r, s, q, m_1, \ldots, m_{n-1}$ are integers with r,s not both zero, $q > 0$, $m_1, \ldots, m_{n-1}$ non-negative and

$$|r|, |s| \leqslant S, \quad q \leqslant Q, \quad m_1 + \ldots + m_{n-1} \leqslant k.$$

From Lemma 1.2 and Lemma 6.4 we see that if $\lambda_i \leqslant L$ and

$\nu_i \leqslant k$ the algebraic numbers

$$\beta_i = \wp((r+s\tau)u_i/q, \lambda_i, \nu_i) \qquad (0 \leqslant i \leqslant n-1)$$

have denominators b_i, independent of λ_i and ν_i, not exceeding $c_2^{k(S+Q)^2}$. Hence if a_i is the least denominator for α_i $(1 \leqslant i \leqslant n-1)$ it follows that bR is an algebraic integer, where

$$b = (a_1 \ldots a_{n-1})^k b_0 \ldots b_{n-1}.$$

Now $a_i \leqslant H$ and

$$H < e^{2h}, \quad |\alpha_i| \leqslant dH, \quad \binom{m_i}{\mu_i} \leqslant 2^{m_i}$$

so that

$$b < c_3^{hk + k(S+Q)^2}$$

and the size of bR does not exceed

$$b2^{m_1+\ldots+m_{n-1}} (dH)^{(n-1)k} B^n \qquad (77)$$

where B is an upper bound for the size of β_i. But clearly from Lemma 1.2 and Lemma 6.4 again

$$B < k^{c_4 k} c_5^{k(S+Q)^2},$$

and so the size of bR is at most $c_6^{hk+k(S+Q)^2}$. We reserve this estimate in its full generality for later, and for the moment we take $S = h$, $Q = 1$, so that the upper bound is $c_7^{h^2k}$.

The algebraic number bR lies in the field $\mathcal{M}$ generated over $\mathbb{Q}$ by the numbers

$$\alpha_1, \ldots, \alpha_{n-1}, \tau, g_2, g_3, \wp(u_i), \wp'(u_i), \wp''(u_i) \qquad (0 \leqslant i \leqslant n-1)$$

and $[\mathcal{M}:\mathbb{Q}] = m < c_8$. Therefore from Lemma 1.6 there is an integral basis $w_1, \ldots, w_m$ of $\mathcal{M}$ with the size of w_i at most H^{c_9}, and we may write

$$bR = n_1w_1 + \dots + n_mw_m$$

where $n_1, \dots ,n_m$ are rational integers with absolute values at most $c_{10}^{h^2k}$. The requirements of Lemma 7.5 will therefore be satisfied if the m equations

$$\sum_{\lambda_0=0}^{L} \dots \sum_{\lambda_{n-1}=0}^{L} \sum_{\mu_1=0}^{m_1} \dots \sum_{\mu_{n-1}=0}^{m_{n-1}} p(\lambda_0, \dots ,\lambda_{n-1})n_i = 0 \qquad (1 \leqslant i \leqslant m)$$

hold for all integers $r,s,m_1, \dots ,m_{n-1}$ in the specified ranges. The total number of equations for the $p(\lambda_0, \dots ,\lambda_{n-1})$ is

$$M \leqslant 5mh^2(k + 1)^{n-1} < k^{n-5/8}$$

while the total number of unknowns is

$$N = (L + 1)^n > k^{n-1/4}.$$

Since

$$M/(N - M) < h^{-1}$$

we see from Lemma 1.7 that the $p(\lambda_0, \dots ,\lambda_{n-1})$ may be chosen as rational integers, not all zero, with absolute values at most

$$(Nc_{10}^{h^2k})^{h^{-1}} < c_1^{hk},$$

and this completes the proof of the lemma.

Lemma 7.6

Let $m_1, \dots ,m_{n-1}$ be non-negative integers with $m_1 + \dots + m_{n-1} \leqslant k$, and let r,s,q be integers with r,s not both zero, $q > 0$ and

$$|r|,|s| \leqslant S , \quad q \leqslant Q ,$$

and suppose that

$$Q \leqslant k^{8n} \leqslant S \leqslant \exp (k^{1/16}).$$

Then

$$|u_1^{m_1} \dots u_{n-1}^{m_{n-1}} A - \Phi| < e^{-\frac{1}{2}H^\varepsilon}$$

where

$$A = A(m_1, \dots, m_{n-1}, r, s, q), \quad \Phi = \Phi(m_1, \dots, m_{n-1}, r, s, q).$$

Furthermore, if $A \neq 0$ then

$$|A| > c_{11}^{-kS^2 Q^{2n}}. \qquad (77a)$$

<u>Proof</u>

If we write

$g = g((r+s\tau)/q, \dots, (r+s\tau)/q) = (\alpha_1 u_1 + \dots + \alpha_{n-1} u_{n-1})(r+s\tau)/q$ and $g_0 = (r+s\tau)u_0/q$ then for integers r,s,q satisfying the conditions of the lemma it is clear from (75) that

$$|g - g_0| < c_{12}\, e^{k^{1/16}}\, e^{-H^\varepsilon} < e^{-\frac{3}{4}H^\varepsilon},$$

since $H > e^{k^{1/8}}$. Let δ denote the distance from g_0 to the nearest pole of $\wp(z)$; from Lemma 6.4 the absolute value of $\wp(g_0)$ is at most $c_{13}^{S^2}$ and thus $\delta > c_{14}^{-S^2}$. Also we have

$$\wp(g,\lambda,\nu) - \wp(g_0,\lambda,\nu) = \frac{\nu!(g-g_0)}{(2\pi i)^2} \int_{C'} \int_{C} \frac{(\wp(z))^\lambda \, dz\, dz'}{(z-z')^{\nu+1}(z'-g)(z'-g_0)} \qquad (78)$$

where C is the positively described circle in the z-plane centred at z' with radius $\frac{1}{4}\delta$, and C' is the positively described circle in the z'-plane centred at g_0 with radius $\frac{1}{4}\delta$. For clearly C' contains no pole of $\wp(z')$ but does contain the point g, since

$$|g - g_0| < e^{-\frac{3}{4}H^\varepsilon} < \tfrac{1}{8}c_{14}^{-S^2} \leqslant \delta/8.$$

Also C contains no poles of $\wp(z)$, for if z is inside C we have

$$|z - \Omega| \geqslant |g_0 - \Omega| - |z - z'| - |z' - g_0| \geqslant \tfrac{1}{2}\delta. \qquad (79)$$

Hence if $\lambda \leqslant L$, $\nu \leqslant k$ the right side of (78) does not exceed

$$k!e^{-\frac{3}{4}H^{\varepsilon}}(c_{15}\delta^{-2})^{L}(c_{16}\delta)^{-k-1} < e^{-\frac{5}{8}H^{\varepsilon}}$$

in absolute value, for (79) implies that $|\wp(z)| < c_{15}\delta^{-2}$ on C.

Now $u_1^{m_1} \ldots u_{n-1}^{m_{n-1}}A$ and Φ are linear combinations of $\wp(g_0,\lambda,\nu)$ and $\wp(g,\lambda,\nu)$ respectively with the same coefficients. The absolute values of these coefficients are at most $c_{17}^{kS^2}$ from the estimates of the previous lemma; hence

$$|u_1^{m_1} \ldots u_{n-1}^{m_{n-1}} A - \Phi| < c_{18}^{kS^2} e^{-\frac{5}{8}H^{\varepsilon}} < e^{-\frac{1}{2}H^{\varepsilon}},$$

and this proves the first part of the lemma.

For the second part we recall from the preceding lemma that there is a rational integer $b \neq 0$ such that b and bR are algebraic integers of size at most $c_{19}^{kS^2}$. Since b is independent of $\lambda_0, \ldots, \lambda_{n-1}$ it follows that bA is an algebraic integer of similar size; by Lemma 6.4 its degree is at most $c_{20}Q^{2n}$ and thus if $A \neq 0$ the asserted estimate for $|A|$ is an immediate consequence of $|\text{Norm}(bA)| \geq 1$.

For the purposes of extrapolation the lower bound just derived is generally not good enough. Although it will permit some extension of the range of r and s in (76), what we really need is a reduction of the exponent k in (77a). We now use Lemma 6.5 to obtain such an improvement under certain conditions with Q = 1. This lemma asserts that at least $\frac{1}{2}S^2$ of the numbers $r+s\tau$ with $|r|,|s| \leq S$ are such that the size of $\wp((r+s\tau)u_i)$ does not exceed c_{22} for $0 \leq i \leq n-1$. We call such numbers special points, and the process of extrapolation will be confined to these points.

Lemma 7.7

Let S be a number with

$$h \leqslant S \leqslant e^{k^{1/16}}$$

and suppose $r+s\tau$ is a special point with $|r|,|s| \leqslant S$. If $m_1, \ldots, m_{n-1}$ are non-negative integers with $m_1 + \ldots + m_{n-1} \leqslant k$ such that

$$A(\mu_1, \ldots, \mu_{n-1}, r, s, l) = 0$$

for all non-negative integers $\mu_1, \ldots, \mu_{n-1}$ with

$$\mu_1 + \ldots + \mu_{n-1} < m_1 + \ldots + m_{n-1}$$

and $A = A(m_1, \ldots, m_{n-1}, r, s, l)$, then either $A = 0$ or we have

$$|A| > c_{23}^{-LS^2}.$$

Proof

We introduce the function

$$\Xi(z_1, \ldots, z_{n-1}) = \sum_{\lambda_0=0}^{L} \cdots \sum_{\lambda_{n-1}=0}^{L} p(\lambda_0, \ldots, \lambda_{n-1})(\wp((r+s\tau)z_1))^{\lambda_1} \cdots (\wp((r+s\tau)z_{n-1}))^{\lambda_{n-1}}(\wp((r+s\tau)z_0))^{\lambda_0}$$

where

$$z_0 = \alpha_1(z_1 - u_1) + \ldots + \alpha_{n-1}(z_{n-1} - u_{n-1}) + u_0.$$

Then the differentiation of Ξ imitates the differentiation of Φ and for all non-negative integers $\mu_1, \ldots, \mu_{n-1}$ we have

$$\Xi_{\mu_1,\ldots,\mu_{n-1}}(u_1, \ldots, u_{n-1}) = (r+s\tau)^{\mu_1+\ldots+\mu_{n-1}} A(\mu_1, \ldots, \mu_{n-1}, r, s, l), \quad (80)$$

since $z_i = u_i$ for $1 \leqslant i \leqslant n-1$ implies $z_0 = u_0$. Now from Lemma 6.3 there exist coprime polynomials $G(x)$, $H(x)$ with coefficients that are algebraic integers of $\mathbb{F} = \mathbb{K}(g_2, g_3)$

such that

$$\wp((r+s\tau)z) = G(\wp(z))/H(\wp(z)).$$

Hence if we write

$$\Delta(z_1, \ldots, z_{n-1}) = (H(\wp(z_0)))^L \ldots (H(\wp(z_{n-1})))^L$$

it follows that

$$\Psi(z_1, \ldots, z_{n-1}) = \Delta(z_1, \ldots, z_{n-1})\Xi(z_1, \ldots, z_{n-1})$$

is a polynomial in $\wp(z_0), \ldots, \wp(z_{n-1})$ with coefficients that are algebraic integers of $\mathbb{F}$. For integers $m_1, \ldots, m_{n-}$ satisfying the conditions of the lemma we have

$$\Psi_{m_1,\ldots,m_{n-1}} = \sum_{\mu_1=0}^{m_1} \cdots \sum_{\mu_{n-1}=0}^{m_{n-1}} \binom{m_1}{\mu_1} \cdots \binom{m_{n-1}}{\mu_{n-1}} \Delta_{\nu_1,\ldots,\nu_{n-1}} \Xi_{\mu_1,\ldots,\mu_{n-1}} \qquad (81)$$

where $\nu_i = m_i - \mu_i$, and from Lemma 1.2 the left side of this may be expressed as a polynomial in $\wp(z_i)$, $\wp'(z_i)$ and $\wp''(z_i)$ for $0 \leqslant i \leqslant n-1$. Furthermore when we multiply this polynomial by $a = (a_1 \ldots a_{n-1})^k$ its coefficients become algebraic integers of the field $\mathbb{F}(\alpha_1, \ldots, \alpha_{n-1})$. On putting $z_i = u_i$ $(1 \leqslant i \leqslant n-1)$ in (81) all the terms except one vanish from (80) and the hypotheses of the lemma; we conclude that a'A is an algebraic integer of the field $\mathbb{M}$ appearing in the proof of Lemma 7.5, where

$$a' = a(r+s\tau)^k\Delta(u_1, \ldots, u_{n-1})$$

is also an algebraic integer. Also $a' \neq 0$, for $H(\wp(z))$ vanishes only if z is a torsion point; furthermore the size of a' is at most

$$H^{nk}(c_{24}S)^k c_{25}^{LS^2} < c_{26}^{LS^2}$$

on using the estimates of Lemma 6.3 for the coefficients of $G(x)$ and $H(x)$. Now since $r+s\tau$ is a special point it follows from the expressions for R and Lemma 1.2 that the size of A is at most c_{27}^{hk} . Hence if $A \neq 0$ the inequality $|\mathrm{Norm}\ (a'A)| \geq 1$ implies the estimate of the lemma.

Lemma 7.8

Let

$$P(z_1, \ldots, z_{n-1}) = (\sigma(u_1 z_1))^{2L} \ldots (\sigma(u_{n-1} z_{n-1}))^{2L} (\sigma(g))^{2L} .$$

Then

$$\phi(z_1, \ldots, z_{n-1}) = P(z_1, \ldots, z_{n-1})\Phi(z_1, \ldots, z_{n-1})$$

is an entire function of $z_1, \ldots, z_{n-1}$ and for any non-negative integers $m_1, \ldots, m_{n-1}$ with $m_1 + \ldots + m_{n-1} \leq k$ we have

$$|\phi_{m_1,\ldots,m_{n-1}}(z, \ldots, z)| < c_{28}^{hk+L|z|^2} ,$$

$$|P_{m_1,\ldots,m_{n-1}}(z, \ldots, z)| < c_{29}^{hk+L|z|^2} .$$

Proof

It is clear from Lemma 7.1 that ϕ is an entire function. Also if $z, z_1, \ldots, z_{n-1}$ are any complex numbers with $|z_i - z| \leq H^{-1}$ $(1 \leq i \leq n-1)$ then

$$|g(z_1, \ldots, z_{n-1})| < c_{30}|z| + H^{-1} \sum_{i=1}^{n-1} |\alpha_i u_i| < c_{31}(|z| + 1),$$

and thus again from Lemma 7.1

$$|\phi(z_1, \ldots, z_{n-1})| < c_{32}^{hk} c_{33}^{L(|z|+1)^2} < c_{34}^{hk+L|z|^2}$$

with a similar estimate for $|P(z_1, \ldots, z_{n-1})|$. From Cauchy's formula we have

$$\phi_{m_1,\ldots,m_{n-1}}(z, \ldots, z) = \frac{m_1! \ldots m_{n-1}!}{(2\pi i)^{n-1}} \int_{C_1} \ldots \int_{C_{n-1}} \frac{\phi(z_1, \ldots, z_{n-1})\, dz_1 \ldots dz_{n-1}}{\prod_{i=1}^{n-1} (z_i - z)^{m_i+1}}$$

where C_i is the positively described circle with centre

at z and radius H^{-1} ; thus we obtain

$$|\phi_{m_1,\dots,m_{n-1}}(z, \dots ,z)| \leqslant k!H^{nk}c_{34}^{hk+L|z|^2} < c_{28}^{hk+L|z|^2}$$

The estimates for the derivatives of P follow in exactly the same way, and this proves the lemma.

Now that the basic estimates are established, we employ the techniques of extrapolation. Our aim is to show that for a certain $q > 1$ the range of r and s in the equations

$$A(m_1, \dots ,m_{n-1},r,s,q) = 0 \qquad (82)$$

can be significantly extended at the expense of slightly restricting the range of $m_1, \dots ,m_{n-1}$. This will be achieved in two stages: in the next lemma we obtain this assertion for special $r+s\tau$ and $q = 1$, and then by a process of interpolation we include the required division values.

Lemma 7.9

Let J be an integer with

$$0 \leqslant J \leqslant 2k^{1/64n} .$$

Then (76) holds for all non-negative integers $m_1, \dots ,m_{n-1}$ with

$$m_1 + \dots + m_{n-1} \leqslant k - Jk^{1-1/32n}$$

and all special points $r+s\tau$ with

$$|r|,|s| \leqslant hk^{J/32n} .$$

Proof

The lemma is valid for $J = 0$ by Lemma 7.5. We write for brevity $t = k^{1/64n}$ and we let I be an integer with $0 \leqslant I < 2t$; further we assume that the lemma holds for

$0 \leqslant J \leqslant I$. We proceed to establish its validity for $J = I + 1$.

We write

$$\kappa = [k/t^2] \quad , \quad S_J = ht^{2J} \quad , \quad T_J = k - J\kappa$$

and we suppose the lemma to be false for $J = I + 1$; thus there exist non-negative integers $m_1', \ldots, m_{n-1}'$ with

$$m_1' + \ldots + m_{n-1}' \leqslant T_{I+1} \quad ,$$

and a special point $\rho' = r' + s'\tau$ with

$$|r'|, |s'| \leqslant S_{I+1}$$

such that

$$A' = A(m_1', \ldots, m_{n-1}', r', s', l) \neq 0, \tag{83}$$

and we shall deduce a contradiction from the additional hypothesis that $m_1', \ldots, m_{n-1}'$ are chosen minimally in the usual way. In the notation of Lemma 7.8 we set

$$\psi(z) = \phi_{m_1', \ldots, m_{n-1}'}(z, \ldots, z),$$

whence for integers $m \geqslant 0$, r, s with r,s not both zero we have

$$\psi_m(r+s\tau) = \sum_{\substack{\mu_1=0 \\ \mu_1+\ldots+\mu_{n-1}=m}}^{m} \ldots \sum_{\mu_{n-1}=0}^{m} m!(\mu_1! \ldots \mu_{n-1}!)^{-1} \phi_{\nu_1, \ldots, \nu_{n-1}}(r+s\tau, \ldots, r+s\tau) \tag{84}$$

where $\nu_i = m_i' + \mu_i$. The derivatives of ϕ on the right of (84) may be expressed as

$$\sum_{\rho_1=0}^{\nu_1} \ldots \sum_{\rho_{n-1}=0}^{\nu_{n-1}} \binom{\nu_1}{\rho_1} \ldots \binom{\nu_{n-1}}{\rho_{n-1}} P_{\rho_1, \ldots, \rho_{n-1}} \Phi'' \tag{85}$$

where

$$P_{\rho_1, \ldots, \rho_{n-1}} = P_{\rho_1, \ldots, \rho_{n-1}}(r+s\tau, \ldots, r+s\tau)$$

and

$$\Phi'' = \Phi(\nu_1 - \rho_1, \ldots, \nu_{n-1} - \rho_{n-1}, r, s, l).$$

Suppose now that $r+s\tau$ is a special point with $|r|,|s| \leqslant S_I$, and that $m \leqslant \kappa$. Then the order of the derivative of Φ'' does not exceed $T_{I+1} + \kappa = T_I$, and it follows from the induction hypothesis and Lemma 7.6 that $|\Phi''| < e^{-\frac{1}{2}H^{\varepsilon}}$. From Lemma 7.8 we have

$$|P_{f_1,\ldots,f_{n-1}}| < c_{35}^{LS_I^2}$$

and therefore from (84) and (85) we find the basic estimates

$$|\psi_m(r+s\tau)| < k^{c_{36}k} c_{35}^{LS_I^2} e^{-\frac{1}{2}H^{\varepsilon}} < e^{-\frac{1}{4}H^{\varepsilon}}, \qquad (86)$$

since

$$LS_I^2 < Lh^2k^{t/8n} < \exp(k^{1/32}) \quad , \quad \exp(k^{1/8}) < H.$$

We now use the Hermite extrapolation formula to deduce from the inequalities (86) an upper bound for $|\psi(\rho')|$. We write for brevity

$$F(z) = \Pi\ (z - r - s\tau)^{\kappa}$$

where the product is taken over all special points $r+s\tau$ with $|r|,|s| \leqslant S_I$. Then

$$\frac{\psi(\rho')}{F(\rho')} = \frac{1}{2\pi i}\int_C \frac{\psi(z)\ dz}{(z-\rho')F(z)} - \frac{1}{2\pi i}\sum \frac{\psi_m(r+s\tau)}{m!} I(m,r,s) \qquad (87)$$

where C is the positively described circle with centre the origin and radius

$$R = 25\, S_{I+1}\, (1 + |\tau|);$$

$I(m,r,s)$ is given by

$$I(m,r,s) = \int_{C_{rs}} \frac{(z-r-s\tau)^m}{(z-\rho')F(z)}\, dz \qquad (88)$$

where C_{rs} is the positively described circle with centre $r+s\tau$ and radius k^{-6}, and the summation in (87) is over the ranges

$$0 \leqslant m \leqslant \kappa\ , \quad |r|,|s| \leqslant S_I$$

such that r+sτ is special.

We start by estimating the integral (88). Let N_I be the total number of linear factors in the product F(z), so that from Lemma 6.5

$$c_{37}\kappa S_I^{\,2} < N_I < c_{38}\kappa S_I^{\,2}.$$

If z is on the circle C_{rs} we have

$$|z - r - s\tau| = k^{-6}, \quad |F(z)|^{-1} \leqslant k^{6N_I}$$

and since $\rho' \neq r+s\tau$

$$|z - \rho'| \geqslant |\rho' - r - s\tau| - k^{-6} > c_{39}.$$

It follows that

$$|I(m,r,s)| < k^{7N_I} < \exp(H^{\varepsilon}/16).$$

Suppose now that z lies on C. Then from Lemma 7.8

$$|\psi(z)| < c_{40}^{\;LR^2} < 2^{N_I}$$

by virtue of the inequalities

$$LR^2 < c_{41}\,kh^2t^{4I-4}, \quad N_I > c_{42}\,kh^2t^{4I-2}. \qquad (89)$$

Also if r+sτ appears in the product for F(z) we see that

$$|z - r - s\tau| \geqslant 10|\rho' - r - s\tau|$$

whence

$$|F(z)|^{-1} \leqslant 10^{-N_I}|F(\rho')|^{-1}.$$

Finally there is the obvious upper bound

$$|F(\rho')| < (c_{43}S_{I+1})^{N_I} < \exp(H^{\varepsilon}/16),$$

and therefore from (86) and (87) we deduce that

$$|\psi(\rho')| < c_{44}\,2^{N_I}10^{-N_I} + c_{45}N_I e^{-\frac{1}{8}H^{\varepsilon}} < 4^{-N_I}.$$

We now proceed to derive an estimate for the derivative Φ' of Φ corresponding to A'. We have

$$\psi(\rho') = \sum_{\mu_1=0}^{m_1'} \cdots \sum_{\mu_{n-1}=0}^{m_{n-1}'} \binom{m_1'}{\mu_1} \cdots \binom{m_{n-1}'}{\mu_{n-1}} P'_{\mu_1,\dots,\mu_{n-1}}\,\Phi'''$$

where

$$P'_{\mu_1,\dots,\mu_{n-1}} = P_{\mu_1,\dots,\mu_{n-1}}(\rho', \dots ,\rho')$$

and

$$\Phi''' = \Phi(m_1'-\mu_1, \dots ,m_{n-1}'-\mu_{n-1},r',s',1).$$

From the minimal choice of $m_1', \dots ,m_{n-1}'$ and Lemma 7.6 we see that if $\mu_i \neq 0$ for some i then $|\Phi'''| < e^{-\frac{1}{2}H^{\varepsilon}}$. Otherwise, by isolating the term Φ' with $\mu_1 = \dots = \mu_{n-1} = 0$ we get the inequality

$$|\Phi'| < 4^{-N_I}|P|^{-1} + e^{-\frac{1}{4}H^{\varepsilon}}$$

where

$$P = P(\rho', \dots ,\rho').$$

Since ρ' is special it is a consequence of Lemma 6.5 that none of the numbers $\rho'u_i$ $(0 \leqslant i \leqslant n-1)$ is nearer than c_{46} to a pole of $\wp(z)$; thus from Lemma 7.1 we see that

$$|\sigma(\rho'u_i)| > c_{47}^{-S_{I+1}^2}$$

and so $|P| > c_{48}^{-LS_{I+1}^2}$. Again from (89) this implies

$$|\Phi'| < 3^{-N_I}.$$

Now Φ' corresponds to A' in the sense of Lemma 7.6, and hence there follows the estimate

$$|A'| < 2^{-N_I}.$$

On the other hand $A' \neq 0$ by supposition, and therefore the induction hypothesis and Lemma 7.7 give the lower bound

$$|A'| > c_{49}^{-LS_{I+1}^2}.$$

These are contradictory and this completes the proof of the lemma.

Lemma 7.10

For all non-negative integers $m_1, \ldots, m_{n-1}$ with

$$m_1 + \ldots + m_{n-1} \leqslant L$$

and all integers r,s not both zero with absolute values at most $\exp(k^{1/128n})$ we have

$$A(m_1, \ldots, m_{n-1}, r, s, q) = 0$$

where $q = k^5$.

Proof

From the preceding lemma (76) holds with $J = [t]$. If the lemma is false there exist integers $m_1', \ldots, m_{n-1}'$, r', s' satisfying the conditions above such that

$$A' = A(m_1', \ldots, m_{n-1}', r', s', q) \neq 0$$

where $m_1', \ldots, m_{n-1}'$ are chosen minimally. We define the function $\psi(z)$ corresponding to $m_1', \ldots, m_{n-1}'$; since $L \leqslant T_{I+1}$ and $I + 1 < 2t$ for $I = [t]$ we can deduce the inequalities (86) as before; furthermore $\rho' = (r'+s'\tau)/q$ cannot be one of the numbers $r+s\tau$ appearing there. Because

$$S_{I+1} \geqslant S = \exp(k^{1/128n})$$

the integral formula (87) holds for ρ' since the radius of C_{rs} is small enough to keep ρ' well away from its interior. Hence by estimating as before we obtain the inequality

$$|\psi(\rho')| < 4^{-N_I}.$$

From the minimal choice of $m_1', \ldots, m_{n-1}'$ and Lemma 7.6 we find that

$$|\Phi'| < 4^{-N_I}|P|^{-1} + e^{-\frac{1}{4}H\varepsilon}$$

where $P = P(\rho', \ldots, \rho')$ and

$$\Phi' = \Phi(m_1', \ldots, m_{n-1}', r', s', q).$$

The bounds for $|\wp(\rho' u_i)|$ given by Lemma 6.4 imply that none of the points $\rho' u_i$ are nearer than $c_{50}^{-S^2}$ to a pole of $\wp(z)$; thus $|P| > c_{51}^{-LS^2}$ and $|\Phi'| < 3^{-N_I}$. Since Φ' corresponds to A' in the sense of Lemma 7.6 we deduce that

$$|A'| < 2^{-N_I}.$$

On the other hand since $A' \neq 0$ a lower bound is provided by the same lemma;

$$|A'| > c_{52}^{-kS^2 q^{2n}}.$$

But $N_I > e^{4t}$ and

$$kS^2 q^{2n} < k c_{53} e^{2t} < e^{3t}$$

whence the contradiction proves the lemma.

7.5 The Wronskian

We proceed to derive the final contradiction from the vanishing of the algebraic numbers in Lemma 7.10. We shall eventually make use of Lemma 6.7, but first we employ a process of elimination similar to the arguments of section 2.5. This has the effect of reducing by one the number of elliptic functions appearing in the auxiliary function, and it enables the theorem to be used inductively.

Let $F(\lambda_{n-1})$ denote the function

$$F(\lambda_{n-1}, z_1, \ldots, z_{n-1}) = \sum_{\lambda_0=0}^{L} \cdots \sum_{\lambda_{n-2}=0}^{L} p(\lambda_0, \ldots, \lambda_{n-1})(\wp(u_1 z_1))^{\lambda_1} \cdots (\wp(u_{n-2} z_{n-2}))^{\lambda_{n-2}} (\wp(g))^{\lambda_0}$$

so that

$$\Phi(z_1, \ldots, z_{n-1}) = \sum_{\lambda_{n-1}=0}^{L} F(\lambda_{n-1})(\wp(u_{n-1}z_{n-1}))^{\lambda_{n-1}}.$$

We denote by F the vector space spanned by the functions $F(0), \ldots, F(L)$ over $\mathbb{C}$, and we shall assume until the closing sentences of this section that the dimension of F over $\mathbb{C}$ is M+1 for $0 \leqslant M \leqslant L$. The contradiction we shall eventually obtain will therefore prove that F is zero. Let $F(r_0), \ldots, F(r_M)$ be a basis for F. Then complex numbers $c(\lambda,\mu)$ $(0 \leqslant \lambda \leqslant L, 0 \leqslant \mu \leqslant M)$ exist such that

$$F(\lambda) = \sum_{\mu=0}^{M} c(\lambda,\mu)F(r_\mu)$$

and we have

$$\Phi = \sum_{\mu=0}^{M} G(\mu)F(r_\mu), \tag{90}$$

where

$$G(\mu) = G(\mu, z_{n-1}) = \sum_{\lambda=0}^{L} c(\lambda,\mu)(\wp(u_{n-1}z_{n-1}))^{\lambda}.$$

We note that $G(\mu)$ is a non-zero polynomial in $\wp(u_{n-1}z_{n-1})$ since $c(r_\mu, \mu) = 1$.

Let $\partial_0, \ldots, \partial_M$ be an arbitrary set of differential operators of the form

$$(\partial/\partial z_1)^{m_1} \ldots (\partial/\partial z_{n-2})^{m_{n-2}}$$

with $|\partial_\mu| \leqslant \mu$ for $0 \leqslant \mu \leqslant M$ (recall that $n-2 \geqslant 1$). Then from (90)

$$\partial_\nu \Phi = \sum_{\mu=0}^{M} G(\mu)\partial_\nu F(r_\mu), \tag{91}$$

and we regard this as a set of linear equations between the vector with components $\partial_\nu \Phi$ $(0 \leqslant \nu \leqslant M)$ and the vector with components $G(\mu)$ $(0 \leqslant \mu \leqslant M)$. The determinant W of these equations is a Wronskian of the functions $F(r_\mu)$ $(0 \leqslant \mu \leqslant M)$

and from Lemma 1.2 we can write it in the form

$$W = W(z_1, \ldots, z_{n-1}) = \sum qX_0^{\nu_0} \ldots X_{n-2}^{\nu_{n-2}} X_0'^{\nu_0'} \ldots X_{n-2}'^{\nu_{n-2}'} X_0''^{\nu_0''} \ldots X_{n-2}''^{\nu_{n-2}''} \qquad (92)$$

where

$$q = q(\nu_0, \ldots, \nu_{n-2}, \nu_0', \ldots, \nu_{n-2}', \nu_0'', \ldots, \nu_{n-2}''),$$

$$X_i = \wp(u_i z_i), \quad X_i' = \wp'(u_i z_i), \quad X_i'' = \wp''(u_i z_i) \quad (1 \leqslant i \leqslant n-2)$$

and

$$X_0 = \wp(g), \quad X_0' = \wp'(g), \quad X_0'' = \wp''(g).$$

The summation is over all indices ν_i, ν_i', ν_i'' satisfying

$$0 \leqslant \nu_i, \nu_i', \nu_i'' \leqslant N = 2L(L+1) \qquad (0 \leqslant i \leqslant n-2)$$

since from Lemma 1.2 each entry of the Wronskian has the form (92) where the indices range from 0 to 2L.

The following device has the effect of removing the variables X_i', X_i'' from W (cf. p.60). Let $\sigma = \{n_1, \ldots, n_r\}$ be an ordered subset of the integers between 0 and n-2, and define the numbers $q^{(\sigma)}$ by

$$q^{(\sigma)} = (-1)^s q$$

where

$$s = \nu_{n_1}' + \ldots + \nu_{n_r}'.$$

Then the function $W^{(\sigma)}$ given by (92) with $q^{(\sigma)}$ in place of q may be regarded as a conjugate of W obtained by changing the sign of X_i' whenever i is in the set σ. In fact there is a closer relation between $W^{(\sigma)}$ and W which may be written in the form

$$W(z_1, \ldots, z_{n-1}) = W^{(\sigma)}(z_1^{(\sigma)}, \ldots, z_{n-1}^{(\sigma)}) \qquad (93)$$

where $z_i^{(\sigma)}$ is a linear form in $z_1, \ldots, z_{n-1}$. For if

$1 \leqslant i \leqslant n-2$ we simply define $z_i^{(\sigma)} = \pm z_i$ and take the negative sign if and only if i is in σ, and to correct the remaining X_0' we define $z_{n-1}^{(\sigma)}$ from the equation

$$g(z_1^{(\sigma)}, \ldots, z_{n-1}^{(\sigma)}) = \pm g(z_1, \ldots, z_{n-1}) \qquad (94)$$

where the negative sign is chosen if and only if 0 is in σ. We note that since $\alpha_{n-1} \neq 0$ (94) genuinely does admit a solution $z_{n-1}^{(\sigma)}$.

We now assert that the function

$$U = U(z_1, \ldots, z_{n-1}) = \prod_{\sigma} W^{(\sigma)}(z_1, \ldots, z_{n-1}), \qquad (95)$$

where σ runs over the 2^{n-1} possible subsets, can be expressed as

$$U = \sum_{\rho_0=0}^{R} \ldots \sum_{\rho_{n-2}=0}^{R} r(\rho_0, \ldots, \rho_{n-2}) X_0^{\rho_0} \ldots X_{n-2}^{\rho_{n-2}}. \qquad (96)$$

Here the coefficients $r(\rho_0, \ldots, \rho_{n-2})$ are certain complex numbers and $R \leqslant k^2$. For, considered as a polynomial in the variables X_i, X_i', X_i'' $(0 \leqslant i \leqslant n-2)$, U is clearly an even function of each X_i'. Hence it is a polynomial in X_i, $X_i'^2$ and X_i'', and the expression (96) follows since $X_i'^2$ and X_i'' are both polynomials in X_i.

Finally we define f(z) as the function obtained from $U(z, \ldots, z)$ by replacing $g(z, \ldots, z)$ in X_0 by $u_0 z$; thus

$$f(z) = \sum_{\rho_0=0}^{R} \ldots \sum_{\rho_{n-2}=0}^{R} r(\rho_0, \ldots, \rho_{n-2})(\wp(u_0 z))^{\rho_0} \ldots (\wp(u_{n-2} z))^{\rho_{n-2}}.$$

<u>Lemma 7.11</u>

For all $x \neq 0$ in Λ/k^5 with absolute value at most $\exp(k^{1/256n})$ we have $f(x) = 0$, with possibly 2L exceptions.

<u>Proof</u>

Let x be a non-zero point of Λ/k^5 with absolute value at most $\exp(k^{1/256n})$, and write

$$\Xi(z_1, \ldots, z_{n-1}) = \sum_{\lambda_0=0}^{L} \ldots \sum_{\lambda_{n-1}=0}^{L} p(\lambda_0, \ldots, \lambda_{n-1})(\wp(xz_1))^{\lambda_1} \ldots (\wp(xz_{n-1}))^{\lambda_{n-1}}(\wp(xz_0))^{\lambda_0}$$

where

$$z_0 = \alpha_1(z_1 - u_1) + \ldots + \alpha_{n-1}(z_{n-1} - u_{n-1}) + u_0$$

(cf. proof of Lemma 7.7). Corresponding to (90) we get a similar decomposition of Ξ, and by applying the differential operators $\partial_0, \ldots, \partial_M$ to both sides we obtain a set of linear equations like (91) between the vector with components $\partial_\nu \Xi$ and the vector with components $R(\mu)$, where

$$R(\mu) = R(\mu, z_{n-1}) = \sum_{\lambda=0}^{L} c(\lambda,\mu)(\wp(xz_{n-1}))^{\lambda}.$$

Furthermore, since the differentiation of Ξ mimics the differentiation of Φ (cf. the identity (80)), the determinant of this linear system is, apart from a multiplicative factor of powers of $u_1, \ldots, u_{n-1}$ and x, the expression (92) with $X_i = \wp(xz_i)$, $X_i' = \wp'(xz_i)$, $X_i'' = \wp''(xz_i)$ for $0 \leqslant i \leqslant n-2$.

Now if we put $z_i = u_i$ $(1 \leqslant i \leqslant n-1)$ the quantities $\partial_\nu \Xi$ all vanish, since the integers r,s defined by $x = (r+s\tau)/q$ satisfy the conditions of Lemma 7.10 and

$$\Xi_{\mu_1,\ldots,\mu_{n-1}}(u_1, \ldots, u_{n-1}) = x^{\mu_1+\ldots+\mu_{n-1}} A(\mu_1, \ldots, \mu_{n-1}, r, s, q).$$

Hence if the vector with components $R(\mu, u_{n-1})$ does not vanish, the determinant must be zero for $z_i = u_i$ $(1 \leqslant i \leqslant n-1)$. But, for example, the first component $R(0, u_{n-1})$ of this vector is a non-zero polynomial of degree at most L in $\xi = \wp(xu_{n-1})$ and so vanishes for at most L values of ξ;

since u_{n-1} is not a torsion point there correspond to each such value of ξ at most two values of x. Therefore (92) vanishes with $X_i = \wp(xu_i)$ $(0 \leqslant i \leqslant n-2)$, with at most 2L exceptions. Clearly f(x) is the product of this expression with its conjugates, and this proves the lemma.

We now use the results of section 7.2 to complete the proof. We begin by showing that the numbers $r(\rho_0, \ldots, \rho_{n-2})$ vanish for all $\rho_0, \ldots, \rho_{n-2}$. Let $\phi(X_0, \ldots, X_{n-2})$ denote the polynomial (96), and let $\mathcal{B}$ denote the unit ball

$$|X_0|^2 + \ldots + |X_{n-2}|^2 \leqslant 1$$

in $\mathbb{C}^{n-1}$. Let $(\xi_0, \ldots, \xi_{n-2})$ be an arbitrary point of $\mathcal{B}$. By our basic induction hypothesis, Theorem V is valid for the n-1 algebraic points $u_0, \ldots, u_{n-2}$. Hence if $x_0, \ldots, x_{n-2}$ are lattice points of Λ, not all zero, with absolute values at most k^5, their heights are at most k^{11}, and on taking $\varepsilon = (400n)^{-1}/11$ in the theorem we deduce that

$$|u_0 x_0 + \ldots + u_{n-2} x_{n-2}| > \exp(-k^{1/300n}).$$

Thus from Lemma 6.7 there are at least k distinct points x of Λ/k^5 such that

$$|\eta_i - \xi_i| < k^{-3} \qquad (0 \leqslant i \leqslant n-2)$$

where $\eta_i = \wp(u_i x)$, and furthermore these points satisfy the conditions of Lemma 7.11. Hence we must have f(x) = 0 for at least one of these points, and since $f(x) = \phi(\eta_0, \ldots, \eta_{n-2})$ it follows that the polynomial ϕ has a zero within $\sqrt{n-1}.k^{-3}$ of $(\xi_0, \ldots, \xi_{n-2})$. Since

$$\sqrt{n-1}.k^{-3} < (2(n-1)^2 R)^{-1}$$

we deduce from Lemma 7.4 that ϕ is identically zero. This implies that all the coefficients $r(\rho_0, \ldots, \rho_{n-2})$ vanish.

Thus from (95) and (96) one of the conjugates $W^{(\sigma)}(z_1, \ldots, z_{n-1})$ is identically zero, and therefore by the identities (93) the Wronskian $W(z_1, \ldots, z_{n-1})$ must be identically zero. Since the differential operators $\partial_0, \ldots, \partial_M$ were arbitrarily selected, every Wronskian of the functions $F(r_\mu)$ $(0 \leqslant \mu \leqslant M)$ not involving $\partial/\partial z_{n-1}$ vanishes identically. Lemma 7.3 now asserts the existence of functions $H(\mu) = H(\mu, z_{n-1})$, independent of $z_1, \ldots, z_{n-2}$ and not all zero, such that

$$\sum_{\mu=0}^{M} H(\mu) F(r_\mu) = 0. \tag{97}$$

The same lemma also gives an expression for $H(\mu)$ as a rational function of total degree at most M in the elements of the Wronskian matrices involved. Hence we conclude that $H(\mu)$ is a rational function of total degree at most $c_{54} L^2$ in the functions

$$\wp(u_1 z_1), \ldots, \wp(u_{n-2} z_{n-2}), \quad \wp(\alpha_1 u_1 z_1 + \ldots + \alpha_{n-1} u_{n-1} z_{n-1})$$

and their first and second derivatives. Lemma 7.2 was constructed expressly for this situation. Since either α_1 is not in $\mathbb{K}$ or its height is $H > \exp(k^{1/8}) > L^3$ by the suppositions of the Proposition, it follows that $H(\mu)$ is in fact independent of z_{n-1} $(0 \leqslant \mu \leqslant M)$, and now the linear relation (97) contradicts the linear independence of $F(r_0), \ldots, F(r_M)$ over the complex numbers.

Returning now to the beginning of this section we conclude from this contradiction that the vector space F contains no non-zero elements. Hence the functions $F(0), \ldots, F(L)$ vanish identically. Since $\alpha_{n-1} \neq 0$, it is clear that the elliptic functions appearing in $F(\lambda_{n-1})$ are algebraically independent, whence $p(\lambda_0, \ldots, \lambda_{n-1})$ vanishes for all $\lambda_0, \ldots, \lambda_{n-1}$. This contradicts the choice of these coefficients made in Lemma 7.5, and the final contradiction completes the proof of Theorem V.

7.6 The Case n = 2; A Postscript

The above methods fail when n = 2 because we cannot eliminate one of the elliptic functions by differentiation. But it is easy to indicate an argument that does work. The auxiliary function is

$$\Phi(z_1) = \sum_{\lambda_0=0}^{L} \sum_{\lambda_1=0}^{L} p(\lambda_0,\lambda_1)(\wp(u_1z_1))^{\lambda_1}(\wp(\alpha_1u_1z_1))^{\lambda_0}$$

and by extrapolation we have, for example

$$|\Phi(z_1)| < e^{-k^{10}}$$

for all z_1 with $|z_1| \leqslant k^3$ such that u_1z_1/ω_1 and $\alpha_1u_1z_1/\omega_1$ lie in the set $\mathcal{E}$ of Chapter I. To use these inequalities we find distinct $\lambda_0, \ldots, \lambda_L$ in Λ such that the conditions are fulfilled by the numbers $z_1(\ell,m)$ defined by

$$u_1z_1(\ell,m)/\omega_1 = \tfrac{1}{4} + \ell L^{-2} + \lambda_m \qquad (0 \leqslant \ell,m \leqslant L).$$

In fact, each λ_m must satisfy

$$||\lambda_m\alpha_1 + (\alpha_1 - 1)/4|| < c_1$$

for some small c_1. Now we can use the transference theorem to solve

$$\|\lambda_0\alpha_1 + (\alpha_1 - 1)/4\| < \tfrac{1}{2}c_1 , \quad |\lambda_0| < X ,$$

with

$$X^{-1} = c_2 \min_{0<|\mu|<c_3} \|\mu\alpha_1\| ,$$

and by replacing α_1 by u_0/u_1 we see that $X \leqslant c_4$. We can now take

$$\lambda_m = \lambda_0 + m\lambda$$

where $\lambda \neq 0$ in Λ is a solution of

$$\|\lambda\alpha_1\| < L^{-2} , \quad |\lambda| < c_5L^2$$

For a double application of Lemma 1.3 we require a lower bound for

$$\|\alpha_1(\lambda_m - \lambda_n)\| \qquad (m \neq n).$$

But the conditions on α_1 appearing in the Proposition (see 7.3) imply that this cannot vanish, and the consequent lower bound H^{-c_6} suffices to obtain estimates for $p(\lambda_0,\lambda_1)$ which supply the final contradiction.

APPENDIX ONE

In this appendix we define a function $\psi(z)$ on the upper half $\mathcal{H}$ of the complex plane that is invariant under all transformations of the modular group Γ acting on z. We shall obtain an arithmetical property of this function connected with the theory of complex multiplication which sharpens and illuminates Lemma 3.1.

We use the well known modular forms (see [24])

$$E_4(z) = 1 + 240 \sum_{n=1}^{\infty} n^3 q^n (1 - q^n)^{-1},$$

$$E_6(z) = 1 - 504 \sum_{n=1}^{\infty} n^5 q^n (1 - q^n)^{-1}$$

where $q = e^{2\pi i z}$ and z lies in $\mathcal{H}$; these are related to the modular function

$$j(z) = q^{-1} + 744 + \ldots$$

by

$$j(z) = 1728\, E_4^3/(E_4^3 - E_6^2). \tag{98}$$

Furthermore

$$E_k(-1/z) = z^k E_k(z) \qquad (k = 4,6), \tag{99}$$

and it is well known that the only zeros of $E_6(z)$ are at the points equivalent under Γ to i. If z is not one of these points we define

$$\psi(z) = \frac{3E_4(z)}{2E_6(z)} (E_2(z) - 3/(\pi \mathrm{Im}\, z)) \tag{100}$$

(cf. proof of Lemma 3.2). It is easily checked from (39) that $\psi(z)$ is unaltered by a modular transformation of Γ, and it is only the non-analytic nature of $\psi(z)$ that saves it from being a mere rational function of $j(z)$.

On the other hand, we have the following result*.

Theorem A1

If τ is a complex quadratic irrationality not equivalent to i then $\psi(\tau)$ lies in the field $\mathbb{J}$ generated over $\mathbb{Q}$ by $j(\tau)$.

The proof of this theorem depends on some properties of the modular equation relating $j(z)$ to $j(\underline{\underline{A}}z)$ for an integral matrix $\underline{\underline{A}}$, and we recall the construction of this equation. If two matrices $\underline{\underline{B}}_1$, $\underline{\underline{B}}_2$ are such that $\underline{\underline{B}}_1\underline{\underline{B}}_2^{-1}$ lies in Γ, we say that $\underline{\underline{B}}_1$ and $\underline{\underline{B}}_2$ are j-equivalent, and then it is clear that $j(\underline{\underline{B}}_1 z) = j(\underline{\underline{B}}_2 z)$ for all z for which either side is defined. A primitive matrix is one whose elements are coprime integers; it is well known that for a given integer $D > 1$ there are only finitely many j-equivalence classes of primitive matrices of determinant D. Let $\underline{\underline{C}}_1, \ldots ,\underline{\underline{C}}_n$ denote a complete set of representatives from these classes.

---

* The function $\psi(z)$ does appear in the literature, although not so frequently as its analytic cousin $E_2(z)$ and I cannot find any results on its arithmetical properties. However, after writing this appendix I came across the note [30] of Siegel, in which he uses an approach similar to the present one to investigate the values of $j'(z)$ at complex quadratic numbers.

Then there is a polynomial $\Phi(X,Y)$ with rational coefficients such that ([27] p.239)

$$\Phi(j(z),Y) = \prod_{i=1}^{n} (Y - J_i(z)) \tag{101}$$

where

$$J_i(z) = j(\underline{\underline{C}}_i z).$$

Furthermore $\Phi(X,Y)$ is symmetric in X and Y, and clearly

$$\Phi(j(z),j(\underline{\underline{A}}z)) = 0$$

for all primitive matrices $\underline{\underline{A}}$ of determinant D.

For a complex quadratic irrationality τ not equivalent to i we fix D as follows. There exist coprime integers A,B,C > 0 such that

$$A + B\tau + C\tau^2 = 0.$$

We denote by $\underline{\underline{A}}$ the matrix

$$\begin{pmatrix} -B & -2A \\ 2C & B \end{pmatrix} \quad \text{or} \quad \begin{pmatrix} -B/2 & -A \\ C & B/2 \end{pmatrix}$$

according to whether B is odd or even, and we let $D = \det \underline{\underline{A}}$. Since τ is not equivalent to i we have D > 1, otherwise B would be even and the quadratic forms $Ax^2 + Bxy + Cy^2$, $x^2 + y^2$ would be equivalent. In the next two lemmas we determine the leading terms in the Taylor expansion of $\Phi(X,Y)$ about the point

$$X = Y = j = j(\tau).$$

It is convenient to single out a certain class of τ for special treatment; we call τ special if B is odd and $D = 3d^2$ for an integer d > 1.

Lemma A1

Let $\underline{\underline{B}}$ be a primitive matrix of determinant D such that $\tau = \underline{\underline{B}}\tau$. Then up to j-equivalence $\underline{\underline{B}} = \underline{\underline{A}}$ unless τ is special, in which case $\underline{\underline{B}} = \underline{\underline{A}}$ or

$$\underline{\underline{B}} = \underline{\underline{B}}^{\pm} = \begin{pmatrix} \frac{1}{2}(-B\pm 3d) & -A \\ C & \frac{1}{2}(B\pm 3d) \end{pmatrix}$$

Proof

Assuming that $\underline{\underline{B}}$ satisfies the conditions of the lemma and setting

$$\underline{\underline{B}} = \begin{pmatrix} P & Q \\ R & S \end{pmatrix}$$

we deduce that coprime integers r,s exist with

$$R = Cr \ , \quad S = Br + s \ , \quad P = s \ , \quad Q = -Ar \ .$$

Now the determinant condition implies that if B is odd

$$(2s + Br)^2 = D(4 - r^2) \qquad (102)$$

while if B is even

$$(2s + Br)^2 = 4D(1 - r^2). \qquad (103)$$

Since $D > 1$ the solution $r = 0$ does not yield a primitive matrix $\underline{\underline{B}}$; also $r = \pm 2$ in (102) or $r = \pm 1$ in (103) give the original matrix $\underline{\underline{A}}$. There remains the solution $r = \pm 1$ in (102) and this implies $D = 3d^2$ and $\underline{\underline{B}} = \underline{\underline{B}}^{\pm}$. Finally it is straightforward but tedious to check that $\underline{\underline{A}}$, $\underline{\underline{B}}^{+}$, $\underline{\underline{B}}^{-}$ are all j-inequivalent if and only if $d > 1$, and if $d = 1$ they are all j-equivalent.

Lemma A2

The expansion of $\Phi(X,Y)$ about the point $X = Y = j$ is

given by

$$\Phi(X,Y) = \beta(X - j) + \beta(Y - j) + \ldots \qquad (\beta \neq 0)$$

unless τ is special, in which case

$$\Phi(X,Y) = \beta(X - j)^3 + \beta(Y - j)^3 + \ldots \qquad (\beta \neq 0)$$

where the missing terms are of higher order.

Proof

We write

$$\Phi(X,Y) = \sum_{\mu} \sum_{\nu} \beta_{\mu\nu}(X - j)^{\mu}(Y - j)^{\nu} \qquad (\mu,\nu \geqslant 0)$$

where the $\beta_{\mu\nu}$ clearly lie in $\mathbb{J}$. We take $\underline{\underline{C}}_1 = \underline{\underline{A}}$ in (101) and put $z = \tau$; this gives the well-known result

$$\beta_{00} = \Phi(j,j) = 0.$$

Further, by differentiating (101) with respect to Y and putting $z = \tau$ we find that

$$\beta_{01} = \prod_{i=2}^{n} (j(\tau) - J_i(\tau))$$

whence $\beta_{01} \neq 0$ if τ is not special. For if $j(\tau) = J_i(\tau)$ with $i \geqslant 2$ we would have $\tau = \underline{\underline{E}}\underline{\underline{C}}_i\tau$ for some $\underline{\underline{E}}$ in Γ and from Lemma A1 it would follow that $\underline{\underline{C}}_i$ and $\underline{\underline{A}} = \underline{\underline{C}}_1$ are j-equivalent. From symmetry $\beta_{01} = \beta_{10}$, and this proves the lemma if τ is not special.

If τ is special the following expansions near $z = \tau$ may readily be verified:

$$j(z) = j + (z - \tau)j'(\tau) + \tfrac{1}{2}(z - \tau)^2 j''(\tau) + \ldots ,$$

$$j(\underline{\underline{A}}z) = j - (z - \tau)j'(\tau) + \tfrac{1}{2}(z - \tau)^2 J_1''(\tau) + \ldots , \qquad (104)$$

$$j(\underline{\underline{B}}^{\pm}z) = j + \tfrac{1}{2}(1 \mp \sqrt{-3})(z - \tau)j'(\tau) + \ldots$$

and we take $\underline{\underline{C}}_2 = \underline{\underline{B}}^{+}$, $\underline{\underline{C}}_3 = \underline{\underline{B}}^{-}$ in (101). Then by substituting the above expansions into the formulae

$\Phi(j(z),J_i(z)) = 0$ $(i = 1,2,3)$ we see that $\beta_{\mu\nu} = 0$ for $\mu+\nu < 3$, and also that the numbers $\theta = -1, \frac{1}{2}(1 \pm \sqrt{-3})$ satisfy the cubic equation

$$\beta_{30} + \beta_{21}\theta + \beta_{12}\theta^2 + \beta_{03}\theta^3 = 0.$$

Hence $\beta_{21} = \beta_{12} = 0$, and finally by differentiating (101) three times with respect to Y and putting $z = \tau$ we conclude that

$$\beta_{03} = 6 \prod_{i=4}^{n} (j(\tau) - J_i(\tau)) \neq 0$$

again from Lemma A1.

The proof of the theorem is completed as follows. We substitute the first two expansions of (104) into the identity $\Phi(j(z),J_1(z)) = 0$. On equating to zero the coefficients of $(z - \tau)^2$ or $(z - \tau)^4$ according to whether τ is not special or is special, we find that

$$j''(\tau) + J_1''(\tau) = -2(j'(\tau))^2\gamma$$

where

$$\gamma = (\beta_{20} - \beta_{11} + \beta_{02})/\beta$$

unless τ is special, when

$$\gamma = (\beta_{40} - \beta_{31} + \beta_{22} - \beta_{13} + \beta_{04})/\beta.$$

Since

$$J_1''(\tau) = j''(\tau) + 4Cj'(\tau)/(B + 2C\tau)$$

we obtain

$$j''(\tau)/j'(\tau) + 2C/(B + 2C\tau) = -j'(\tau)\gamma. \qquad (105)$$

It remains to express the left side of (105) in terms of $\psi(\tau)$. For this we use the identities

$$E_2'(z) = \tfrac{1}{6}\pi i(E_2^2 - E_4),$$

$$E_4'(z) = \tfrac{2}{3}\pi i(E_2E_4 - E_6), \quad E_6'(z) = \pi i(E_2E_6 - E_4^2)$$

which can be verified by the usual weight arguments. From (98) we get

$$j'(z) = -2\pi i E_6 j(z)/E_4$$

and thus

$$j''(z)/j'(z) = \tfrac{1}{3}\pi i E_2 - \tfrac{4}{3}\pi i E_6/E_4 - \pi i E_4^2/E_6,$$

whence from (105) and the observation

$$(B + 2C\tau)/2C = i \operatorname{Im} \tau$$

we see that

$$E_2(\tau) - 3/(\pi \operatorname{Im} \tau) = 4E_6/E_4 + 3E_4^2/E_6 + 6j\gamma E_6/E_4.$$

We conclude that

$$\psi(\tau) = 9j\gamma + 3(7j - 6912)/2(j - 1728) \qquad (106)$$

and this proves Theorem A1.

From the well-known formulae

$$E_4(\tau) = \tfrac{3}{4}g_2(\omega_1/\pi)^4 , \quad E_6(\tau) = \tfrac{27}{8}g_3(\omega_1/\pi)^6$$

it is clear that if τ is not equivalent to either i or $\rho = \frac{1}{2}(-1 + \sqrt{-3})$ the algebraic number κ of Lemma 3.1 is given by

$$\kappa = -(B + 2C\tau)g_3\psi(\tau)/g_2. \qquad (107)$$

Hence the theorem gives a more precise formulation of this lemma.

Although (106) provides a theoretical method of exactly determining $\psi(\tau)$ for a given τ, in practice the calculation of the modular polynomial $\Phi(X,Y)$ is extremely laborious. We end this appendix with a simple technique more suitable for numerical computation. For an integer $\lambda \neq 0$ of $\mathbb{K}$ we define the λ-division points of $\wp(z)$ as the points u such that λu

but not u is a pole of $\wp(z)$; then the λ-division values of $\wp(z)$ are the values of $\wp(u)$ as u runs over all mutually incongruent λ-division points. It is not hard to verify that the number of such incongruent points is Norm $\lambda - 1$.

Lemma A3

If τ is not equivalent to i or ρ we have

$$\psi(\tau) = -g_2 S/Cg_3(2A + B\tau)$$

where S is the sum of the $C\tau$-division values of $\wp(z)$.

Proof

We consider the function

$$g(z) = f(z) - (1 - A/C)\zeta(z) + (A/C)\sum_u \zeta(z-u) - \sum_v \zeta(z-v)$$

where f(z) is the function (37), and u, v run over all C-division points and $C\tau$-division points respectively up to congruence. Since the residue of $\zeta(z)$ at any pole is unity it is easily seen that g(z) has no poles. But f(z) is an elliptic function and hence

$$\begin{aligned} g(z+\omega_i) - g(z) &= -(1 - A/C)\eta_i + (A/C)(C^2 - 1)\eta_i - (AC - 1)\eta_i \\ &= 0 \qquad (i = 1,2), \end{aligned}$$

since there are AC - 1 values of v. It follows that g(z) is a constant. We now note that near z = 0

$$\zeta(z+w) = \zeta(w) - \wp(w)z + O(z^2) \qquad (w \neq \text{pole})$$

$$\zeta(z) = z^{-1} + O(z^3)$$

and we equate to zero the coefficient of z in the Laurent expansion of g(z). We find that

$$C\tau\kappa = (A/C)\sum_u \wp(u) - \sum_v \wp(v).$$

But from Lemma 6.1 the $\wp(u)$ are the totality of roots of the polynomial $B_\ell(x)$ for $\ell = C$, and it is easy to verify by induction on the functions $\psi_\ell(x)$ that the second highest term in this polynomial is absent. Hence the sum over u vanishes, and now the assertion of the lemma follows from (107).

To determine $\psi(\tau)$ we proceed as follows. We choose g_2, g_3 such that $\frac{1}{4}g_2$, $\frac{1}{4}g_3$ are algebraic integers; then by Lemma 4 of [2] the numbers $(AC)^2\wp(v)$ are algebraic integers and we can easily find a denominator for $\psi(\tau)$. The size of $\psi(\tau)$ can be estimated in a similar way, and since Theorem A1 (or Lemma 3.1) gives an explicit field containing $\psi(\tau)$, it only remains to compute $\psi(\tau)$ numerically to an arbitrary degree of accuracy. This is best done by expressing (100) as a rapidly convergent q-series.

In this way we have evaluated the value of $\psi(\tau)$ for the twelve values of τ in the fundamental region $\mathcal{R}$ of p.51 for which $\psi(\tau)$ is defined and $j(\tau)$ is rational.

τ	$\psi(\tau)$
$\sqrt{-2}$	15/28
$\sqrt{-3}$	15/22
$\frac{1}{2}(-1+\sqrt{-3})$	0
$\sqrt{-4}$	11/14
$\sqrt{-7}$	255/266
$\frac{1}{2}(-1+\sqrt{-7})$	5/14
$\frac{1}{2}(-1+\sqrt{-11})$	48/77
$\frac{1}{2}(-1+\sqrt{-19})$	16/19

τ	$\psi(\tau)$
$\frac{1}{2}(-1 + \sqrt{-27})$	240/253
$\frac{1}{2}(-1 + \sqrt{-43})$	320/301
$\frac{1}{2}(-1 + \sqrt{-67})$	16720/14539
$\frac{1}{2}(-1 + \sqrt{-163})$	38632640/30285563

APPENDIX TWO

We investigate here some problems raised by Lemma 2.3 and Lemma 7.4. Let

$$\phi(\underline{z}) = \sum_{\lambda_1=0}^{L} \cdots \sum_{\lambda_n=0}^{L} p(\lambda_1, \ldots, \lambda_n) z_1^{\lambda_1} \ldots z_n^{\lambda_n}$$

be a non-zero polynomial of degree at most L in each variable, and denote by $\mathcal{B}_R$ the set of real points of the unit ball $\mathcal{B}$ given by $|\underline{z}| \leqslant 1$. From the above lemmas we know that constants c_1, c_2, depending only on n, exist such that ϕ cannot have a zero within c_1L^{-2} of every point of $\mathcal{B}_R$ or within c_2L^{-1} of every point of $\mathcal{B}$. (The extension of Lemma 2.3 to n variables is perfectly straightforward using (73).) On the basis of simple counting arguments it is reasonable to conjecture that the best possible results with regard to dependence on L would be with c_3L^{-1} and $c_4L^{-1/2}$ respectively.

In this appendix we use elementary methods to prove a version of Lemma 2.3 with $c_5(L \log L)^{-1}$ in place of c_1L^{-2}; this does not quite establish the conjecture. However, we show that the conjecture for $\mathcal{B}$ is a simple consequence of a result of Lelong on the measure associated with a holomorphic function on $\mathbb{C}^n$. We also include a corresponding refinement of this conjecture giving an upper bound for the coefficients $p(\lambda_1, \ldots, \lambda_n)$ in terms of the values of ϕ on a suitably dense set. Such a refinement does not seem to be a straightforward application of Lelong's ideas.

We start with the result for $\mathcal{B}_R$, which may be summarized in the following theorem.

Theorem A2

There is a constant c_5 depending only on n such that ϕ cannot have a zero within $c_5(L \log L)^{-1}$ of every point of $\mathcal{B}_R$.

For the proof we require a lemma concerning polynomial approximations to the function $(\log z)/2\pi i$. We write for an integer $m \geqslant 1$

$$q_m(z) = -(2\pi i)^{-1} \sum_{r=1}^{m} (1 - z)^r/r. \tag{108}$$

Lemma A4

If m is larger than an absolute constant then for any complex number a with $|a| < 1/40$ there is a complex number $\zeta = \zeta(m,a)$ such that

$$q_m(\zeta) = a \;, \quad |\zeta - e^{2\pi i a}| < 2^{-m}.$$

Proof

We first derive a lower bound for the absolute value of $(\log z)/2\pi i$ on the circle C with centre at $z = 1$ and radius $\frac{1}{2}$, where the branch of $\log z$ is chosen to coincide with the limit of the series (108). Writing for brevity

$$g(z) = \sum_{r=2}^{\infty} (1 - z)^r/r$$

we have on C the obvious inequalities

$$|g(z)| \leq g(\tfrac{1}{2}) = \log 2 - \tfrac{1}{2} < 1/5 ,$$

so that

$$|\log z| \geq |1 - z| - |g(z)| \geq 3/10 .$$

Hence if $|a| \leq 1/40$ we see that on C

$$|a - (\log z)/2\pi i| \geq 3/20\pi - 1/40 > 1/80 .$$

Also if m is sufficiently large we have inside C

$$|(2\pi i)^{-1} \sum_{r=m+1}^{\infty} (1 - z)^r/r| < (m2^m)^{-1} . \qquad (109)$$

Therefore by Rouché's theorem the functions $a - q_m(z)$ and $a - (\log z)/2\pi i$ have the same number of zeros inside C; if $|a| \leq 1/40$ it is easily verified that

$$|e^{2\pi i a} - 1| < \tfrac{1}{2}$$

whence this number is precisely one. If ζ denotes the zero of $a - q_m(z)$ inside C we have from (109)

$$|a - (\log \zeta)/2\pi i| < (m2^m)^{-1} ,$$

and then the inequality

$$|e^x - e^y| \leq e^{|x|+|x-y|} |x - y| \qquad (110)$$

shows that

$$|\zeta - e^{2\pi i a}| < c_6 (m2^m)^{-1} < 2^{-m}$$

which proves the lemma.

To deduce the theorem suppose that ϕ has a zero within $\delta < c_7$ of each point of $\mathcal{B}_R$. Let m be any integer with

$$2 \log \delta^{-1} < m < 3 \log \delta^{-1} ,$$

and let

$$\Phi(z_1, \ldots ,z_n) = \phi(q_m(z_1\sqrt{n}), \ldots ,q_m(z_n\sqrt{n})).$$

Suppose $\underline{x} = (x_1, \ldots ,x_n)$ is any point of $\mathcal{B}_R$ with $|x_j| \leq 1/80$ $(1 \leq j \leq n)$. Let $\underline{a}$ be the zero of ϕ nearest to $\underline{x}$, so that

$$|x_j - a_j| \leq \delta \quad , \quad |a_j| \leq 1/40 \qquad (1 \leq j \leq n).$$

From the above lemma there exist complex numbers $\zeta_1, \ldots ,\zeta_n$ such that

$$q_m(\zeta_j) = a_j \quad , \quad |\zeta_j - e^{2\pi i a_j}| < 2^{-m} < \delta \qquad (1 \leq j \leq n) \quad ,$$

and from (110) we see that

$$|e^{2\pi i x_j} - e^{2\pi i a_j}| < c_8 \delta$$

whence

$$|\zeta_j - e^{2\pi i x_j}| < c_9 \delta$$

It follows that the polynomial $\Phi(z_1, \ldots, z_n)$ has a zero $(\zeta_1/\sqrt{n}, \ldots, \zeta_n/\sqrt{n})$ within $c_{10}\delta$ of every point of the region $\mathfrak{U}$ (see p.98) whose coordinates have arguments at most $2\pi/80$ in absolute value. Therefore the product

$$\Psi(z_1, \ldots, z_n) = \Pi \; \Phi(\varepsilon_1 z_1, \ldots, \varepsilon_n z_n)$$

has a zero within $c_{10}\delta$ of every point of $\mathfrak{U}$ if the numbers $\varepsilon_1, \ldots, \varepsilon_n$ range independently over all 80-th roots of unity. This corresponds to the weaker hypothesis of Lemma 7.4, and since the degree of Ψ in each variable is at most $c_{11}\, mL$, we conclude that

$$c_{10}\delta > c_{12}\,(mL)^{-1} > c_{13}\,(L \log \delta^{-1})^{-1}$$

or

$$\delta > c_5 (L \log L)^{-1} .$$

This completes the proof. It is easy to obtain a corresponding refinement in the manner of Lemma 2.3.

For the results on the set $\mathfrak{B}$, we recall the basic properties of the measure associated with a holomorphic function $f(z_1, \ldots, z_n)$. This is a positive measure μ on $\mathbb{C}^n$ given by

$$\mu(\mathcal{E}) = \int_{\mathcal{E} \cap \mathfrak{W}_f} \chi(\mathcal{E}) \; d\sigma$$

where $\chi(\mathcal{E})$ is the characteristic function of the set $\mathcal{E}$, $\mathfrak{W}_f$ is the divisor of zeros of f, and $d\sigma$ is the element of area on $\mathfrak{W}_f$. If $\mathfrak{B}(\underline{a},r)$ denotes the ball $|\underline{z} - \underline{a}| \leq r$ we define the function

$$\Theta(\underline{a},r) = \gamma_n r^{2-2n} \mu(\mathcal{B}(\underline{a},r))$$

where $\gamma_n = (n-1)!/\pi^{n-1}$, and we note the following properties of Θ.

(i) $\Theta(\underline{a},r)$ is monotone non-decreasing in r.

(ii) $\Theta(\underline{a}) = \lim_{r\to 0} \Theta(\underline{a},r)$ is the order of the zero of f at $\underline{a}$, so that $\Theta(\underline{a}) = 0$ if and only if $f(\underline{a}) \neq 0$.

Lastly, if f is a polynomial:

(iii) $\lim_{r\to\infty} \Theta(\underline{a},r)$ is independent of $\underline{a}$ and equals the degree d(f) of f, i.e. the greatest value of $\lambda_1 + \dots + \lambda_n$ for which the coefficient $p(\lambda_1, \dots, \lambda_n)$ is non-zero.

Proofs and references for these assertions may be found in the paper [8] of Bombieri.

Suppose $\mathcal{S} \subseteq \mathcal{B}$ contains N points. Then we define the separation δ of $\mathcal{S}$ to be the minimum of 1 and the minimum distance between distinct points of $\mathcal{S}$, and we define the spread of $\mathcal{S}$ as $N\delta^{2n-2}$. We shall prove the following theorem.

<u>Theorem A3</u>

There is a function $c(n) > 0$ of n such that if $\mathcal{S} \subseteq \mathcal{B}$ has separation δ and spread s at least c(n)L we have for all $\lambda_1, \dots, \lambda_n$

$$|p(\lambda_1, \dots, \lambda_n)| \leq (c_{14} L/\delta)^{nL} \max_{z \text{ in } \mathcal{S}} |\phi(\underline{z})|.$$

To see that this implies the conjecture for $\mathcal{B}$ in the form stated above, suppose that ϕ has a zero within $\varepsilon < c_{15}$ of every point of $\mathcal{B}$. We let $N = (4\varepsilon)^{-1}$ and for each set of integers μ_i, ν_i with

$$|\mu_i| \leqslant N/4n \ , \quad |\nu_i| \leqslant N/4n \qquad (1 \leqslant i \leqslant n)$$

we let $\underline{\sigma}$ be the zero of ϕ nearest the point $\underline{p}$ in $\mathcal{B}$ given by

$$\underline{p} = ((\mu_1 + i\nu_1)/N, \ \ldots \ ,(\mu_n + i\nu_n)/N).$$

Let $\mathcal{S}$ be the set of $\underline{\sigma}$ obtained in this way. If $\underline{\sigma}_1$, $\underline{\sigma}_2$ are nearest $\underline{p}_1$, $\underline{p}_2$ respectively and $\underline{p}_1 \neq \underline{p}_2$ we have

$$|\underline{\sigma}_1 - \underline{\sigma}_2| \geqslant |\underline{p}_1 - \underline{p}_2| - |\underline{\sigma}_1 - \underline{p}_1| - |\underline{\sigma}_2 - \underline{p}_2| \geqslant 2\varepsilon,$$

and in particular $\underline{\sigma}_1 \neq \underline{\sigma}_2$ so that $\mathcal{S}$ contains at least $c_{16}\,\varepsilon^{-2n}$ distinct points; it follows that the spread of $\mathcal{S}$ exceeds $c_{17}\varepsilon^{-2}$. Since ϕ vanishes on $\mathcal{S}$ the theorem now implies that $\varepsilon > c_{18}\,L^{-1/2}$.

We start with three lemmas.

Lemma A5

Let π denote the projection from $\mathbb{C}^n$ to $\mathbb{C}^{n-1}$ defined for $n \geqslant 2$ by

$$\pi(z_1, \ \ldots \ ,z_n) = (z_1, \ \ldots \ ,z_{n-1}).$$

Then if $\mathcal{S} \subseteq \mathcal{B}$ has separation δ and spread s, there is a subset $\mathcal{S}'$ of $\pi\mathcal{S}$ with separation δ' and spread s', where

$$\delta' \geqslant \delta \ , \quad s' \geqslant c_{18}\, s.$$

Proof

Let $\mathcal{S}'$ be a maximal subset of $\pi\mathcal{S}$ with separation at least δ, and let s', δ' be the spread and separation respectively of $\mathcal{S}'$. For each $\underline{\sigma}$ in $\mathcal{S}$ there is a $\underline{\sigma}'$ in $\mathcal{S}'$ such that

$$|\underline{\sigma}' - \pi\underline{\sigma}| < \delta \tag{111}$$

otherwise we could adjoin $\pi\underline{\sigma}$ to $\mathcal{S}'$. But for each $\underline{\sigma}'$ in $\mathcal{S}'$ the number of points $\underline{\sigma}$ of $\mathcal{S}$ satisfying (111) is at most $c_{19}\delta^{-2}$ from geometrical considerations; hence the number N' of points in $\mathcal{S}'$ must exceed $c_{20}\delta^2 N$. Therefore

$$s' = N'\delta'^{2n-4} \geq c_{20} N\delta^{2n-2} = c_{20} s$$

and this proves the lemma.

Lemma A6

The spread of any subset of $\mathcal{W}_\phi \cap \mathcal{B}(\underline{0},\frac{3}{2})$ is at most $c_{21} L$.

Proof

Let $\mathcal{S}$ be any subset of $\mathcal{W}_\phi \cap \mathcal{B}(\underline{0},\frac{3}{2})$ with separation δ and spread s. Then for $\underline{\sigma}$ in $\mathcal{S}$ the balls $\mathcal{B}(\underline{\sigma},\frac{1}{2}\delta)$ are disjoint, and from properties (i) and (ii) their measures are at least

$$\gamma_n^{-1}(\tfrac{1}{2}\delta)^{2n-2}\Theta(\underline{\sigma},\tfrac{1}{2}\delta) \geq \gamma_n^{-1}(\tfrac{1}{2}\delta)^{2n-2}\Theta(\underline{\sigma}) \geq \gamma_n^{-1}(\tfrac{1}{2}\delta)^{2n-2}.$$

They are all contained in $\mathcal{B}(\underline{0},2)$ whose measure is at most

$$\gamma_n^{-1} 2^{2n-2}\Theta(\underline{0},2) \leq \gamma_n^{-1} 2^{2n-2} d(\phi)$$

by (i) and (iii). Hence the number of points of $\mathcal{S}$ is at most $(4/\delta)^{2n-2} d(\phi)$, and since $d(\phi) \leq nL$ the lemma follows.

Lemma A7

For an integer $L \geq 1$ and complex numbers $p(\lambda)$ $(0 \leq \lambda \leq L)$ let

$$f(z) = \sum_{\lambda=0}^{L} p(\lambda) z^\lambda$$

be a polynomial of degree at most L that does not vanish inside the unit disc $|z| \leq 1$. Then for all λ we have

$$|p(\lambda)| \leq (c_{22} L)^L |f(0)|.$$

Proof

By the maximum modulus principle applied to $1/f(z)$ there is a point σ_λ on the circle $|z| = \lambda/L$ $(0 \leq \lambda \leq L)$ such that $|f(\sigma_\lambda)| \leqslant |f(0)|$. The lemma now follows easily from Lemma 1.3, since for $\lambda \neq \mu$ $|\sigma_\lambda - \sigma_\mu| \geqslant 1/L$.

We proceed to prove Theorem A3 by induction on n. For $n = 1$ it is a simple consequence of Lemma 1.3; hence we assume its validity for $n-1$ variables and deduce it for n variables.

For a point $\underline{\sigma} = (\sigma_1, \ldots, \sigma_n)$ of $\mathcal{S}$ we write

$$\psi(z_n) = \psi(\underline{\sigma}, z_n) = \phi(\sigma_1, \ldots, \sigma_{n-1}, z_n)$$

and we define the coefficients $q(\lambda_n)$ by

$$\psi(z_n) = \sum_{\lambda_n=0}^{L} q(\lambda_n)((z_n - \sigma_n)/\tfrac{1}{4}\delta)^{\lambda_n}. \tag{112}$$

Let $\mathcal{S}_1$ be the subset of $\mathcal{S}$ consisting of the points $\underline{\sigma}$ such that $\psi(\underline{\sigma}, \sigma_n') = 0$ for some σ_n' with $|\sigma_n' - \sigma_n| \leqslant \tfrac{1}{4}\delta$; for $\underline{\sigma}$ in $\mathcal{S}_1$ we write $\underline{\sigma}' = (\sigma_1, \ldots, \sigma_{n-1}, \sigma_n')$, and we denote by $\mathcal{S}'$ the set of such points $\underline{\sigma}'$. Then since $|\underline{\sigma}' - \underline{\sigma}| \leqslant \tfrac{1}{4}\delta$ the separation δ' of $\mathcal{S}'$ is at least $\tfrac{1}{2}\delta$ and from Lemma A6 the number of points N' in $\mathcal{S}'$ satisfies

$$N'\delta'^{2n-2} < c_{21} L.$$

If N is the number of points of $\mathcal{S}$ this gives

$$N' < c_{23} L\delta^{2-2n} \leqslant c_{24} NL/s \leqslant c_{24} N/c(n) \;;$$

hence if $c(n)$ is sufficiently large there must be a subset $\mathcal{S}_2$ of $\mathcal{S}$ containing at least $\tfrac{1}{2}N$ points not in $\mathcal{S}_1$. For $\underline{\sigma}$ in $\mathcal{S}_2$ the polynomial $\psi(z_n)$ does not vanish on the disc $|z_n - \sigma_n| \leqslant \tfrac{1}{4}\delta$ and from Lemma A7 and (112) we deduce that

$$|q(\lambda_n)| \leqslant (c_{25} L)^L |\psi(\sigma_n)| \qquad (0 \leqslant \lambda_n \leqslant L). \tag{113}$$

Thus if $\chi(\lambda_n) = \chi(\lambda_n, z_1, \ldots, z_{n-1})$ denotes the polynomial

$$\chi(\lambda_n) = \sum_{\lambda_1=0}^{L} \cdots \sum_{\lambda_{n-1}=0}^{L} p(\lambda_1, \ldots, \lambda_n) z_1^{\lambda_1} \cdots z_{n-1}^{\lambda_{n-1}}$$

we have

$$\psi(z_n) = \sum_{\lambda_n=0}^{L} \chi(\lambda_n, \sigma_1, \ldots, \sigma_{n-1}) z_n^{\lambda_n},$$

and, comparing with (112),

$$\chi(\lambda_n, \sigma_1, \ldots, \sigma_{n-1}) = \sum_{\mu=\lambda_n}^{L} \binom{\mu}{\mu-\lambda_n} q(\mu)(-\sigma_n)^{\mu-\lambda_n}(4/\delta)^{\mu}$$

whence, from (113)

$$|\chi(\lambda_n, \sigma_1, \ldots, \sigma_{n-1})| \leqslant (c_{26} L/\delta)^L |\psi(\sigma_n)| \qquad (0 \leqslant \lambda_n \leqslant L).$$

This gives a set of inequalities for $\chi(\lambda_n)$ on the points of $\pi\mathcal{S}_2$. The separation of $\mathcal{S}_2$ is at least δ, and thus its spread is at least $\frac{1}{2}s$; hence from Lemma A5 there is a subset of $\pi\mathcal{S}_2$ with separation at least δ and spread at least $c_{27}\, s$. Hence if $c_{27}\, c(n) \geqslant c(n-1)$ we may apply the theorem to the polynomial $\chi(\lambda_n)$ on this subset, and since $\psi(\sigma_n) = \phi(\underline{\sigma})$ we deduce that for all $\lambda_1, \ldots, \lambda_{n-1}$

$$|p(\lambda_1, \ldots, \lambda_n)| \leqslant (c_{28} L/\delta)^{(n-1)L} (c_{26} L/\delta)^L \max_{\underline{z} \text{ in } \mathcal{S}} |\phi(\underline{\sigma})| \quad (0 \leqslant \lambda_n \leqslant L)$$

and this completes the proof of the theorem.

APPENDIX THREE

In Chapter VII it was shown that algebraic points of a Weierstrass elliptic function with complex multiplication over $\mathbb{K}\,(\neq \mathbb{Q})$ are linearly independent over $\mathbb{A}$ provided they are linearly independent over $\mathbb{K}$. In this appendix we indicate the proof of a companion result, namely that any linear combination of algebraic points with algebraic coefficients is either zero or transcendental. We shall avoid repetition of the details of Chapter VII by restricting ourselves to an outline of the ideas involved.

We need only one preliminary lemma, which bears some resemblance to Lemma 2.1.

Lemma A8

For meromorphic functions $F(\lambda,z)$ $(0 \leqslant \lambda \leqslant L)$ and complex numbers α, β let

$$f(z) = \sum_{\lambda=0}^{L} F(\lambda,z)(\alpha z + \beta)^{\lambda}.$$

Then the Wronskian of the functions

$$f(\mu,z) = \sum_{\lambda=0}^{L-\mu} \binom{\mu+\lambda}{\mu} F(\lambda+\mu,z)(\alpha z + \beta)^{\lambda}$$

is given by

$$W(z) = \det_{0 \leqslant \lambda,\mu \leqslant L} F(\lambda,\mu,z)$$

where the functions $F(\lambda,\mu,z)$ are defined by

$$F(\lambda,0,z) = F(\lambda,z) \qquad (0 \leqslant \lambda \leqslant L),$$

$$F(\lambda,\mu+1,z) = (d/dz)F(\lambda,\mu,z) + \alpha(\lambda+1)F(\lambda+1,\mu,z) \qquad (0 \leqslant \lambda,\mu < L),$$

$$F(L,\mu+1,z) = (d/dz)F(L,\mu,z) \qquad (0 \leqslant \mu < L).$$

Proof

There is a proof in [15], but it can be condensed as follows. Defining the binomial coefficients $\binom{a}{b}$ as zero if $a < b$ it is easy to verify by induction on μ that

$$(d/dz)^{\mu} f(\nu,z) = \sum_{\lambda=0}^{L} \binom{\lambda}{\nu} F(\lambda,\mu,z)(\alpha z + \beta)^{\lambda-\nu}$$

and hence

$$W(z) = \det_{\mu,\lambda} F(\lambda,\mu,z) \quad \det_{\lambda,\nu} \binom{\lambda}{\nu}(\alpha z + \beta)^{\lambda-\nu} .$$

The second factor on the right is a triangular determinant whose main diagonal consists of ones; this proves the lemma.

Contrary to the result we are trying to prove, we suppose that non-zero algebraic numbers $\alpha_0, \ldots ,\alpha_{n-1}$ exist with

$$\alpha_1 u_1 + \ldots + \alpha_{n-1} u_{n-1} = \alpha_0, \tag{114}$$

where without loss of generality $u_1, \ldots ,u_{n-1}$ are algebraic points of $\wp(z)$ linearly independent over $\mathbb{K}$. Also we have seen that if $n \geqslant 3$ we can suppose that u_i $(1 \leqslant i \leqslant n-1)$ is a non-torsion algebraic point, and the case $n = 2$ was settled by Schneider [25].

For a large integer k we set

$$L = [k^{1-1/8n}] \quad , \quad h = [k^{1/8}]$$

and

$$\Phi(z_1, \ldots ,z_{n-1}) = \sum_{\lambda_0=0}^{L} \ldots \sum_{\lambda_{n-1}=0}^{L} p(\lambda_0, \ldots ,\lambda_{n-1})(g(z_1, \ldots ,z_{n-1}))^{\lambda_0} (\wp(u_1 z_1))^{\lambda_1} \ldots (\wp(u_{n-1} z_{n-1}))^{\lambda_{n-1}}$$

where

$$g(z_1, \ldots, z_{n-1}) = \alpha_1 u_1 z_1 + \ldots + \alpha_{n-1} u_{n-1} z_{n-1}.$$

Then from (114) we see that for non-negative integers $m_1, \ldots, m_{n-1}$ and integers r, s, q with $q > 0$, r,s not both zero the number

$$A(m_1, \ldots, m_{n-1}, r, s, q) = u_1^{-m_1} \ldots u_{n-1}^{-m_{n-1}} \Phi_{m_1,\ldots,m_{n-1}}((r+s\tau)/q, \ldots, (r+s\tau)/q)$$

is an algebraic number if the coefficients $p(\lambda_0, \ldots, \lambda_{n-1})$ are rational integers. It is now easy to verify that Lemmas 7.5 to 7.10 remain valid for the new auxiliary function, possibly with modified constants. We proceed as follows. We write

$$F(\lambda_0) = F(\lambda_0, z_1, \ldots, z_{n-1}) = \sum_{\lambda_1=0}^{L} \ldots \sum_{\lambda_{n-1}=0}^{L} p(\lambda_0, \ldots, \lambda_{n-1}) (\wp(u_1 z_1))^{\lambda_1} \ldots (\wp(u_{n-1} z_{n-1}))^{\lambda_{n-1}}$$

and

$$z_0 = \alpha_1 u_1 z_1 + \ldots + \alpha_{n-1} u_{n-1} z_{n-1},$$

so that

$$\Phi(z_1, \ldots, z_{n-1}) = \sum_{\lambda_0=0}^{L} F(\lambda_0) z_0^{\lambda_0}.$$

We assume there exists a maximal integer M with $0 \leqslant M \leqslant L$ such that F(M) is not identically zero, and we set

$$\Phi(\mu) = \Phi(\mu, z_1, \ldots, z_{n-1}) = \sum_{\lambda=0}^{M-\mu} \binom{\mu+\lambda}{\mu} F(\lambda+\mu) z_0^{\lambda} \qquad (0 \leqslant \mu \leqslant M).$$

From Lemma A8 the Wronskian $W(z_1, \ldots, z_{n-1})$ of the functions $\Phi(0), \ldots, \Phi(M)$ with respect to z_1 is given by a determinant which may be written as a polynomial in $\wp(u_1 z_1)$, $\wp'(u_1 z_1)$, $\wp''(u_1 z_1)$ and $\wp(u_i z_i)$ $(2 \leqslant i \leqslant n-1)$ of degree at most $2L(L+1)$ in each variable. Then the function

$U(z_1, \ldots, z_{n-1}) = W(z_1, \ldots, z_{n-1})W(-z_1, z_2, \ldots, z_{n-1})$ is a polynomial in $\wp(u_i z_i)$ $(1 \leq i \leq n-1)$ of degree at most k^2 in each, and from the analogue of Lemma 7.11 we have

$$U(x, \ldots, x) = 0$$

for all $x \neq 0$ in Λ/k^5 with absolute value at most $\exp(k^{1/256n})$, with at most 2L exceptions. As before, this is enough to show that $U(z_1, \ldots, z_{n-1})$, and therefore $W(z_1, \ldots, z_{n-1})$, is identically zero. It follows that functions $H(\mu) = H(\mu, z_2, \ldots, z_{n-1})$, independent of z_1 and not all zero, exist with

$$\sum_{\mu=0}^{M} H(\mu)\Phi(\mu) = 0,$$

and this leads to the equations

$$\sum_{\lambda=0}^{M} \sum_{\mu=0}^{M-\lambda} \binom{\mu+\lambda}{\mu} H(\mu)F(\lambda+\mu)z_0{}^{\lambda} = 0.$$

Since $\alpha_1 u_1 z_1$ is not an algebraic function of $\wp(u_i z_i)$ $(1 \leq i \leq n-1)$ these imply

$$\sum_{\mu=0}^{M-\lambda} \binom{\mu+\lambda}{\mu} H(\mu)F(\lambda+\mu) = 0 \qquad (0 \leq \lambda \leq M).$$

But this is a linear system of equations in $H(0), \ldots, H(M)$ with a triangular determinant whose diagonal consists of the non-zero functions $\binom{M}{\mu}F(M)$. This contradiction proves that all the functions $F(\lambda_0)$ $(0 \leq \lambda_0 \leq L)$ vanish identically, and since the elliptic functions appearing in each are algebraically independent this involves the vanishing of $p(\lambda_0, \ldots, \lambda_{n-1})$ for all $\lambda_0, \ldots, \lambda_{n-1}$. This final contradiction proves the impossibility of (114).

We remark that if $\alpha_0 \neq 0$ a lower bound for the absolute value of the linear form

$$\Lambda = \alpha_0 + \alpha_1 u_1 + \ldots + \alpha_n u_n$$

of the type $|\Lambda| > Ce^{-H^{\varepsilon}}$ may be obtained by elaborating these methods; when taken in conjunction with Theorem V this provides a lower bound for $|\Lambda|$ in complete generality.

APPENDIX FOUR

It is a well-known result of Siegel that any curve of genus 1 has only finitely many integral points, and the effective proof of this given in [5] by Baker and Coates implies slightly more than this, namely, the following. If $F(x,y) = 0$ has genus 1, there is a function $\phi(q)$ with the property that $|x| < \phi(q)$ for all rational points (x,y) with $x = p/q$, $q > 0$ and p,q integers. But the function $\phi(q)$ is very large and the methods of [5], based on investigations into logarithms of algebraic numbers, do not seem suited for a more detailed study. In this appendix we show that Theorem V gives very good bounds for $\phi(q)$ provided we relax our requirements in two ways. First, we must assume the curve we are considering has complex multiplication, and second we must temper our demand for effectiveness. In fact the original theorem of Siegel derived its non-effective nature from two distinct sources, the Thue-Siegel-Roth theorem and the Mordell-Weil basis theorem, and we can eliminate only the first of these. Our result is the following.

Theorem A4

Let $F(x,y) = 0$ be a curve $\mathcal{C}$ of genus 1 defined over $\mathbb{Q}$ with complex multiplication, and let (x,y) be a rational point on $\mathcal{C}$ with $x = p/q$, $q > 0$, p,q integers. Then for any $\varepsilon > 0$ there exists $C > 0$ depending only on $\mathcal{C}$ and ε such that

$$|x| < C \exp((\log q)^{\varepsilon}).$$

Proof

We shall assume $|x| > c$ where c is a large constant depending only on $\mathcal{C}$ and ε. Let $\mathbb{A}(x,y)$ be the function field associated with $\mathcal{C}$. Since $\mathcal{C}$ is birationally equivalent to a curve $\mathcal{C}_W$ in Weierstrass normal form, there exist functions X, Y in $\mathbb{A}(x,y)$ with $\mathbb{A}(x,y) = \mathbb{A}(X,Y)$ and

$$Y^2 = 4X^3 - g_2X - g_3. \tag{115}$$

We let $\mathbb{F}$ be the least field of definition of X, Y, g_2 and g_3.

Suppose now that $x = p/q$ as above. From the equation $F(x,y) = 0$ we deduce that the size and denominator of y do not exceed $(q|x|)^{c_1}$, and it follows that the coordinates X,Y of the corresponding point $\underline{P}$ on $\mathcal{C}_W$ are algebraic numbers of $\mathbb{F}$ with size and denominator at most $(q|x|)^{c_2}$, for if $|x|$ is large enough $\underline{P}$ cannot be a point at infinity on $\mathcal{C}_W$. From the inequality of [20], p.49 we conclude that the height $\exp h(\underline{P})$ of $\underline{P}$ does not exceed $(q|x|)^{c_3}$.

Now the points on $\mathcal{C}_W$ defined over $\mathbb{F}$ form a finitely generated group; let $u_1, \dots, u_r$ be a basis for the points of infinite order under the parametrization by the elliptic function $\wp(z)$ corresponding to (115). If u is the parameter of $\underline{P}$ we may write

$$u = m_1u_1 + \ldots + m_ru_r + n_1\omega_1 + n_2\omega_2 \qquad (116)$$

where $m_1, \ldots, m_r$ are rational integers with absolute values at most M and n_1, n_2 are rational numbers capable of taking only a finite number of values. The Tate height $\hat{h}(\underline{P})$ (see [10], p.262) on $\mathcal{C}_w$ satisfies $|h(\underline{P}) - \hat{h}(\underline{P})| < c_4$ and is a positive definite quadratic form in $m_1, \ldots, m_r$; hence we have $\hat{h}(\underline{P}) > c_5M^2$ and it follows that

$$c_3\log(q|x|) \geqslant h(\underline{P}) \geqslant c_5M^2 - c_4 > c_6M^2,$$

since the inequality $M > c_7$ excludes only a finite number of points (x,y) from consideration. Therefore

$$M < c_8(\log(q|x|))^{1/2} \qquad (117)$$

Now let $\underline{p}_0$ be the point on $\mathcal{C}$ corresponding to an arbitrary valuation on $\mathbb{A}(x,y)$ extending the valuation on $\mathbb{A}(x)$ defined by 1/x, and let

$$X = x^{-m/e}\sum_{k=0}^{\infty}\alpha_k x^{-k/e} \qquad (e > 0\ ,\ \alpha_0 \neq 0) \qquad (118)$$

be the associated Puiseux expansion of X. Taking |x| so large that (118) actually converges, we consider two cases.

(i) $m < 0$. Then we have

$$|X| > c_9|x|^{c_{10}} \qquad (119)$$

while from (116) and Theorem V we see that

$$|u - \ell_1\omega_1 - \ell_2\omega_2| > e^{-M^{\ell/a}}$$

for all integers ℓ_1, ℓ_2, and this gives the upper bound

$$|X| < e^{M^{\varepsilon}}.$$

Together with (117) and (119) this implies that

$$|x| < c_{11} \exp((\log(q|x|))^{1/2})$$

whence the estimate of Theorem A4 follows without difficulty.

(ii) $m \geqslant 0$. Then if $m = 0$ we have

$$|X - \alpha| < c_9 |x|^{-c_{10}}$$

for $\alpha = \alpha_0$ and if $m > 0$ this holds with $\alpha = 0$. Since α is the X-coordinate of the image of $\underline{p}_0$ in $\mathcal{C}_w$ the number v defined by $\wp(v) = \alpha$ is an algebraic point of $\wp(z)$, and we deduce from (116) and Theorem V that

$$|u - v - \ell_1\omega_1 - \ell_2\omega_2| > e^{-M^{\varepsilon/2}}$$

or

$$|X - \alpha| > e^{-M^{\varepsilon}}.$$

The desired inequality follows exactly as in (i) on combining the upper and lower bounds.

In conclusion we observe that the effective determination of the basis $u_1, \ldots, u_r$ becomes possible with the aid of the Birch-Swinnerton-Dyer conjectures. For the analysis of Manin in [22] requires that the zeta-function of $\mathcal{C}$ has a functional equation, and this is well-known to be true if $\mathcal{C}$ has complex multiplication. Therefore if we assume the Birch-Swinnerton-Dyer conjectures the constant C of the theorem becomes effectively computable in terms of the coefficients of the equation defining $\mathcal{C}$.

REFERENCES

[1] BAKER, A. Linear forms in the logarithms of algebraic numbers. Mathematika, 13 (1966), 204 - 216.

[2] BAKER, A. On the periods of the Weierstrass $\wp$-function. Symposia Mathematica, Vol.IV, 155 - 174.

[3] BAKER, A. On the quasi-periods of the Weierstrass ζ-function. Nachr. Akad. Wiss. Göttingen Math.-Phys. Kl.II (1969) 145 - 157.

[4] BAKER, A. An estimate for the $\wp$-function at an algebraic point. Amer. J. Math., 92 (1970), 619 - 622.

[5] BAKER, A. and COATES, J. Integer points on curves of genus 1. Proc. Camb. Phil. Soc. 67 (1970), 595 - 602.

[6] BERNSTEIN, S. Leçons sur les propriétés extrémales et la meilleure approximation des fonctions analytiques d'une variable réelle. Gauthier-Villars, Paris, 1926.

[7] BIRKHOFF, G. and MACLANE, S. A Survey of Modern Algebra. Macmillan, 1965.

[8] BOMBIERI, E. Algebraic values of meromorphic maps. Inventiones Math. 10 (1970), 267 - 287.

[9] CASSELS, J.W.S. An Introduction to Diophantine Approximation. Cambridge Tracts No.45.

[10] CASSELS, J.W.S. Diophantine equations with special reference to elliptic curves. Journal London Math. Soc. 41 (1966), 193 - 291.

[11] COATES, J. An application of the Thue-Siegel-Roth theorem to elliptic functions. Proc. Camb. Phil. Soc. 69 (1971), 157 - 161.

[12] COATES, J. The transcendence of linear forms in ω_1, ω_2, η_1, η_2, $2\pi i$. Amer. J. Math. 93 (1971), 385 - 397.

[13] COATES, J. Linear forms in the periods of the exponential and elliptic functions. Inventiones Math. 12 (1971), 290 - 299.

[14] DIENES, P. The Taylor Series. Oxford 1931.

[15] FELDMAN, N.I. Approximation of certain transcendental numbers II : The approximation of certain numbers associated with the Weierstrass $\wp$-function. Izv. Akad. Nauk. SSSR, Ser. Mat. 15 (1951), 153 - 176. American Math. Translations, Ser.2, Vol.59 (1966), 246 - 270.

[16] FELDMAN, N.I. An elliptic analogue of an inequality of A.O. Gelfond. Trans. Moscow Math. Soc., Vol.18 (1968), 71 - 84.

[17] FRICKE, R. Die elliptische Funktionen und ihre Anwendungen. Vol.II, Leipzig 1916.

[18] GELFOND, A.O. Transcendental and Algebraic Numbers. Dover 1960.

[19] LANDAU, E. Einführung in die elementare und analytische Theorie der algebraischen Zahlen under der Ideale, Leipzig, Teubner, 1918.

[20] LANG, S. Diophantine Geometry. Interscience, New York 1962.

[21] LANG, S. Transcendental Numbers and Diophantine Approximations, Bull. Amer. Math. Soc. 77 (1971), 635 - 677.

[22] MANIN, Ju.I. Cyclotomic fields and modular curves. Russian Math. Surveys, Vol.26, No.6, 7 - 78.

[23] MARKOV, A. Abh. der Akad. der Wiss. zu St. Petersburg, 62, (1889), 1 - 24.

[24] OGG, A. Survey of Modular Functions of One Variable, Modular Functions of One Variable I. Springer-Verlag, 1973.

[25] SCHNEIDER, T. Einführung in die transzendenten Zahlen, Springer, 1957.

[26] SIEGEL, C.L. Uber die Perioden elliptischer Funktionen, J. reine. angew. Math. 167 (1932), 62 - 69.

[27] WEBER, H. Lehrbuch der Algebra, Vol.III, (reprint), Chelsea.

[28] WEYL, H. Algebraic theory of numbers, Ann. of Math. Studies 1, Princeton, 1940.

[29] WHITTAKER, E.T. and WATSON, G.N. Modern Analysis, Cambridge, 1965.

[30] SIEGEL, C.L. Bestimmung der elliptischen Modulfunktion durch eine Transformationsgleichung, Abh. Math. Sem. Univ. Hamburg 27 (1964/1965), 32 - 38.